Lecture Notes in Networks and Systems

Volume 86

The series "Lecture Notes in Networks and Systems" publishes the latest developments in Networks and Systems—quickly, informally and with high quality. Original research reported in proceedings and post-proceedings represents the core of LNNS.

Volumes published in LNNS embrace all aspects and subfields of, as well as new challenges in, Networks and Systems.

The series contains proceedings and edited volumes in systems and networks, spanning the areas of Cyber-Physical Systems, Autonomous Systems, Sensor Networks, Control Systems, Energy Systems, Automotive Systems, Biological Systems, Vehicular Networking and Connected Vehicles, Aerospace Systems, Automation, Manufacturing, Smart Grids, Nonlinear Systems, Power Systems, Robotics, Social Systems, Economic Systems and other. Of particular value to both the contributors and the readership are the short publication timeframe and the world-wide distribution and exposure which enable both a wide and rapid dissemination of research output.

The series covers the theory, applications, and perspectives on the state of the art and future developments relevant to systems and networks, decision making, control, complex processes and related areas, as embedded in the fields of interdisciplinary and applied sciences, engineering, computer science, physics, economics, social, and life sciences, as well as the paradigms and methodologies behind them.

** Indexing: The books of this series are submitted to ISI Proceedings, SCOPUS, Google Scholar and Springerlink **

More information about this series at http://www.springer.com/series/15179

Mikhail V. Belov · Dmitry A. Novikov

Models of Technologies

 Springer

Mikhail V. Belov
IBS Company
Moscow, Russia

Dmitry A. Novikov
V. A. Trapeznikov Institute
of Control Sciences
Russian Academy of Sciences
Moscow, Russia

ISSN 2367-3370   ISSN 2367-3389   (electronic)
Lecture Notes in Networks and Systems
ISBN 978-3-030-31083-7   ISBN 978-3-030-31084-4   (eBook)
https://doi.org/10.1007/978-3-030-31084-4

This Springer imprint is published by the registered company Springer Nature Switzerland AG
The registered company address is: Gewerbestrasse 11, 6330 Cham, Switzerland

Introduction

Without any doubt, the history of mankind development can be called the history of technological progress. Really, technologies are demanded by economy and society; have accelerated growth; are the systemically important (backbone) elements of any production; and finally, maintain the existence and further development of mankind [1–3]. All of these factors determine the conceptual meaning of the above statement. In addition, it seems somewhat populistic: fashionable expressions like "technological revolution," "converging technologies," "neural technologies," "digital technologies," etc. are alternating each other rapidly, causing a gracious smile of professionals and a muddle of men in the street.

In accordance with the definition of the Merriam-Webster Dictionary, technology[1] is (1a) the practical application of knowledge especially in a particular area; (1b): a capability given by the practical application of knowledge; (2) a manner of accomplishing a task especially using technical processes, methods, or knowledge; and (3) the specialized aspects of a particular field of endeavor. This term originates from Greek *technologia* (technē art, skill + -o- + -logia –logy), meaning "systematic treatment of an art."

In [1, 4], a *technology* was defined as a system of conditions, criteria, forms, methods, and means for achieving the desired goal. The models of technology design, adoption, and use described below will rest on this definition.

The models of technologies can be classified in the following general way, in the descending order of their scale:

(1) *"civilization models,"* which reflect the general "macro" laws of technology design and interaction with society over characteristic periods of century or decades (technological structures, Kondratiev cycles, etc. [5–7]);

(2) *"innovations models,"* which study the general laws of innovations initiation, implementation and deployment/diffusion at the micro-level, including the scale of economic sectors and organizations [8] (innovation is a new technology);

---

[1]The term "technology" was introduced in 1772 by German scientist Johann Beckmann to mean the science of trade.

(3) *"activity models,"* which study the general organization laws of any activity, including those of the design and use of different activity technologies [1];
(4) *"models–standards,"* which are being intensively developed in Systems Engineering and contain the well-systematized extensions of best practices from practical or industrial activity [9, 10].
(5) *"subject-matter models,"* which describe specific technologies in different sectors.

This book[2] is focused on the third (activity-related) level of the classification and further develops the original results of the authors presented in [1, 11–13]; also see Chap. 1. A systematic overview of the first two classes of the models seems unreasonable due to their extreme richness and fast evolution; moreover, it would be beyond the scope of this research. The fourth class of the models is fixed, while the fifth one consists of concrete (and specific) elements, and hence, they should not be overviewed too.

Technologies may have different translation forms such as flowcharts and process regulations in industrial production, construction documents in building, network diagrams in project management, and business processes descriptions in the activity of organizations. The general form is an *information model* that describes the actor's states and also the actions (together with the corresponding methods and means) to transform it. Much attention below will be therefore paid to the information models of technologies. At the same time, the computerized design and management tools for the information models of products and technologies known as Continuous Acquisition and Lifecycle Support (CALS) systems—Computer Aided Design, Manufacturing and Engineering (CAD, CAM, CAE) systems and Product Data Management (PDM) systems—will be not considered in this book because they are merely a particular (albeit modern) case of technology translation means.

On the one hand, the design of each technology includes the *general-system* and also *specific* components. We will adopt the general-system approaches only, which neglect any sectoral specifics. On the other hand, the design of each technology includes routine and also *creative* components. This book does not pretend to model creation.

From a mathematical viewpoint, technology is an *algorithm* that describes a multivariant scenario of activity in which multiplicity is caused by external and internal conditions. However, the automatic design and optimization problems of nontrivial algorithms with given properties[3] either cannot be solved in general form or have a very high computational intensiveness. As a result, technology is often designed using its *decomposition* into interconnected simple parts or some *heuristics*.

---

[2]The research was partially supported by the Russian Scientific Foundation, project no. 16-19-10609.

[3]As a rule, the results obtained within the framework of mathematical logic and automata theory are very concrete and can be included in the fifth class of the models (somewhat conventionally).

A technology can be interpreted as a *mapping* of the set of *situations* (current states and, perhaps, the history of system, requirements to result, constraints, etc.) into the set of *actions* and utilized *resources*. In other words, "what, how, and by which means" should be done in a certain situation. As a matter of fact, technology design and adoption consist in proper search and operation of these mappings; see Chap. 2 for details.

A technology is often represented as a *graph*—a finite set of states and transitions between them (perhaps, the latter functionally depend on available resources).

For a technology defined by a *function*, optimization problems can be formulated as follows: Find an optimal value of an efficiency criterion subject to given constraints and properties of the "controlled" system.[4] Such optimization problems will be studied in Chap. 3.

Control mechanisms (the sets of rules and procedures—"mappings") can be treated as a "technology" of managerial decision-making: They describe the desired behavior of a controlled element (agent) and the corresponding decisions of a control element (principal) in different situations. Technologies need to be *optimized*, in many cases using exhaustive and heuristic search methods. The design and adoption of technologies often involve so-called *typical solutions*. Thus, the corresponding analytical complexity and errors have to be analyzed; see Chap. 4 below.

This book is organized[5] as follows. The technology of complex activity and its general models are considered in Chap. 1. The models of technology design and adoption are introduced in Chap. 2. The models of technology management are presented in Chap. 3. Finally, the analytical complexity and errors in solving technology design/optimization problems are estimated in Chap. 4.

References

1. Belov M, Novikov D (2018) Methodology of complex activity. Lenand, Moscow, p 320 (in Russian)
2. Novikov D (2016) Cybernetics: from past to future. Springer, Heidelberg, p 107
3. Schwab K (2016) The fourth industrial revolution. World Economic Forum, Geneva, p 172
4. Novikov A, Novikov D (2007) Methodology. Sinteg, Moscow, p 668 (in Russian)
5. Freeman C, Clark J, Soete L (1982) Unemployment and technical innovation: a study of long waves and economic development. Frances Pinte, London, p 214
6. Lundvall B (ed) (1992) National systems of innovation: towards a theory of innovation and interactive learning. Pinter, London, p 342
7. Nelson R, Winter S (1982) An evolutionary theory of economic change. Harvard University Press, Cambridge, p 454
8. Nonaka I, Takeuchi H (1995) The knowledge-creating company: how japanese companies create the dynamics of innovation. Oxford University Press: Oxford, p 284

---

[4]In accordance with this approach, optimal positional control design is the design and further optimization of control technology.

[5]Chapters consist of sections. Formulas are numbered independently within each chapter, while the figures, tables, examples, and propositions continuously throughout the book.

9. Haskins C (ed) (2012) INCOSE systems engineering handbook version 3.2.2—a guide for life cycle processes and activities. INCOSE, San Diego, p 376

10. MITRE Corporation (2014) Systems engineering guide. MITRE Corporation, Bedford, 2014 p 710

11. Belov M (2018) Theory of complex activity as a tool to analyze and govern an enterprise, proceedings of 13th annual conference on system of systems engineering (SoSE 2018), Paris, pp 541–548

12. Belov M, Novikov D (2017) General-system approach to the development of complex activity technology. Procedia Comput Sci 112:2076–2085.

13. Belov M, Novikov D (2019) Methodological foundations of the digital economy, in big data-driven world: legislation issues and control technologies. Springer, Heidelberg, pp 3–14

Contents

Chapter 1
Technology of Complex Activity

In this chapter, using the results of [1], the technology control problem for *the complex activity*[1] (CA) *of organizational and technical systems* (OTSs) is formalized.

The role and place of technologies in complex activity are discussed in Sect. 1.1. The most important peculiarities of the CA of modern adaptive extended enterprises are analyzed in Sect. 1.2. The formal models of their CA are studied in Sect. 1.3. The information models to manage the technology components of CA are considered in Sect. 1.4. Some well-known models and methods are briefly overviewed in Sect. 1.5. The management problems of technology components of CA are stated in Sect. 1.6.

1.1 Role and Place of Technologies in Complex Activity

Methodology of complex activity. The problematique considered in this subsection is a subset of the control problems for organizational and technical systems (OTSs) and their complex activity (CA), which was thoroughly studied in the monograph [1].

An important result of [1] consists in the fixation of a set of control means for OTSs. Among them, the key role is played by the management of technology components of the CA performed by OTSs. A *technology* is defined as a system of conditions, criteria, forms, methods and means for achieving sequentially a given goal; see the Introduction. *The management of technology components* of CA is understood as the activity towards the development of technology components in the form of information models and their maintenance in an adequate state in accordance with external conditions (environment). This book is focused on the development and

---

[1] An activity is a purposeful behavior of a human. A complex activity is an activity with a nontrivial internal structure, with multiple and/or changing actors, technologies and roles of the subject matter in its relevant context [1].

An organizational and technical system is a complex system that consists of humans, technical and natural elements.

© Springer Nature Switzerland AG 2020
M. V. Belov and D. A. Novikov, *Models of Technologies*, Lecture Notes
in Networks and Systems 86, https://doi.org/10.1007/978-3-030-31084-4_1

management of technology components as major problems. It represents a logical continuation of the monograph [1] but is a separate study as well.

The concept of an *organizational and technical system* used in [1] actually extends the definitions of technical, organizational [2], ergatic and sociotechnical systems, matching in some sense the term *"enterprise"* [3] in the Western academic literature. This term seems more natural for the context of this book. Hereinafter, the concepts of an OTS and enterprise will be used equivalently.

Note that the enterprises themselves create no utility: of crucial importance is their activity, which produces a result of real value. Therefore, while considering enterprises, first of all we have to analyze their complex activity using the approaches and results of *the methodology of complex activity* [1].

The following results were established within the methodology of complex activity [1] and are of direct relevance to technologies.

(1) A technology is a key component of any structural element of activity.
(2) A technology determines the result of CA up to the realized event of uncertainty.
(3) Technology development is the key stage that includes the activity steps for achieving a target result (goal) at the phase of activity implementation. The requirements to this result are formulated and further specified at the stage of goal-setting and structuring of goals and tasks.
(4) Technology design is the activity whose subject matter is always a new (e.g., further specified) information model of CA and/or another subject matter.
(5) The design of new technologies of CA can be described by two processes that are intended to operate information (process (a)) and material objects (process (b)):

   (a) the design and specification of goal, conditions, forms, methods, means (including resources) and criteria;
   (b) the definite organization of material resources (information resources are organized during the design process).

(6) The technology of any element of CA is described by the logical, cause-effect and process models in combination with the models of lower SEAs and elementary operations.
(7) A newly created technology and its result may and must be verified and/or falsified (at the epistemological level) and/or tested at the technology implementation level.
(8) The life cycle of a technology includes three phases as follows: development/creation (an analog of the organization and design phase); productive use, with a possible return to the first phase (an analog of the technological phase or the implementation phase); an infinite existence in form of historical data, with a possible return to the second phase.
(9) Technologies are a link between the animate (staff/OTS) and inanimate (product/technological complex) elements and subject matters of activity.
(10) The need for technology development is a creativeness criterion of activity.

(11) To a considerable degree, technology design is specific and hence can be optimized only up to the development of several alternatives with further choice of a best one.

(12) Managing the technology of CA is a complex activity towards creating the information models of its components (including resources) and maintaining them in actual state through modernizations.

These general postulates will be described in detail below.

Technology and structural element of activity. The monograph [1] laid the theoretical foundations of this research by introducing the methodology of complex activity as an extension of general methodology [4] to the case of any complex human activity with a nontrivial multilevel internal structure. In particular, the basic element of CA modeling and analysis—*the structural element of activity* (SEA)—was identified and the logical, cause-effect (causal) and process structures of complex activity were described in constructive terms.

The model of the structural element of activity is shown in Fig. 1.1 [1]. The arrows in this diagram have the following semantics. The arrow from the actor to the "needs–goal–tasks" aggregate indicates that the actor accepts the demand (and need) and executes the goal-setting. The arrows from the actor to the technology and to the actions indicate that the actor executes actions (acts) in accordance with the technology. The arrow from the result to the actor indicates that the actor evaluates the output as well as performs self-regulation and reflection.

The arrows from the technology and actions to the subject matter indicate that the subject matter is transformed by the action in accordance with the technology. Finally, the arrow from the subject matter to the result indicates that the result is the final state of the transformed subject matter and its evolution in the course of activity.

Technology is the key component of any SEA that determines its result.

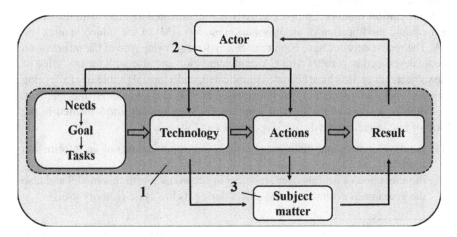

Fig. 1.1 Model of structural element of activity

Table 1.1 Phases, stages and steps of life cycle of CA

Phase	Stage	Step
Design	I. Establishing demand and recognizing needs	1. Establishing demand and recognizing needs
	II. Goal-setting and structuring of goals and tasks	2. Designing logical model
	III. Developing technology	3. Verifying technology readiness and resources sufficiency
		4. Designing cause-effect model
		5. Designing technologies of lower CA elements
		6. Creating/modernizing resources
		7. Scheduling and resource planning
		8. Performing resource optimization
		9. Assigning actors and defining responsibilities
		10. Assigning resources
Implementation	IV. Executing actions and forming output	11. Executing actions and forming output
Reflection	V. Evaluating output and performing reflection	12. Evaluating output and performing reflection

The phases, stages and steps of the life cycle of CA are presented in Table 1.1. Stage III (technology development) is the key stage that includes the activity steps for achieving a target result (goal) at the phase of activity implementation. The requirements to this result are formulated and further specified at the stage of goal-setting and structuring of goals and tasks.

Information models. Complex activity is implemented together with the development and modification of an *information model*[2] (IM) of the subject matter and CA. The recent decades have been remarkable for a growing role of the information model developed in parallel with CA implementation and also with the evolution of the subject matter. IMs have become complicated, and a topical problem is to develop efficient operation procedures for information models and knowledge management.

Complex activity in the field of complex OTSs is actually implemented in the form of two interconnected parallel processes as follows:

(1) the development and maintenance (including modification) of an information model;
(2) the execution of actions over an object in accordance with this model and also the guaranteed evolution of the object during its life cycle (activity itself).

---

[2]We will consider an information model as a model of an object represented in the form of information that describes the significant parameters and variables of the object, the relations between them and also the inputs and outputs of the object. An information model can be used to simulate all possible states of an object by supplying information about its input variations.

This transformation has become an objective source for reviewing the role of information in social life; for example, see numerous discussions on "information explosions," "transition to information society," "digital economy," "knowledge economy," etc.

An information model generally contains not only normative (prior) information on complex activity but also operational (online) information and predictive information for concrete objects as well as various historical data and auxiliary knowledge with different level of detail and degree of formalization.

The gradual complication of IMs and the increase of their role require the development of efficient methods and tools for creating, storing, using, modifying and maintaining IMs. These methods, procedures and means are the subject matter of several fields of knowledge and activity of information technology.

Technology elements. The above definition of technologies underlines that the technology of CA is purposeful, i.e., oriented towards achieving a given goal. This definition not just outlines the set of technology elements but determines their ordered collection, representing a technology as a system. Also this definition specifies the following *technology elements*:

i.   the concrete conditions under which CA is implemented;
ii. the organizational forms of CA;
iii. the methods of CA as a concept that extends operations and techniques;
iv. the implementation means of CA;
v.   the criteria of goal achievement.

All technology elements are informational (knowledge) or material. Conditions (element i) describe in which circumstances and under which rules CA is implemented. In other words, a condition is an information object but it may describe material objects. The same applies to forms (ii), methods (iii) and criteria (v). The implementation means of CA (iv) are, in the first place, resources; see the monograph [1]. Thus, elements i–iii and v are purely informational while element iv is both informational and material. Hence, goal-setting and the development of new technologies of CA can be represented by two processes as follows:

(a) the design and further specification of the goal, conditions, forms, methods, means (including resources) and criteria of CA;
(b) the organization of material resources (note that information resources are organized in the course of design).

Process (a) is intended to operate information while process (b) material objects.

An information model contains all information on the relevant activity. Therefore, we may claim that goal-setting and technology development are an activity with a new (e.g., detailed) information model as the subject matter. In particular, it establishes a correspondence between elements i–iii and resources. In special cases, the resources themselves can be the subject matter too.

A prerequisite for goal-setting and technology development is to predict how technology implementation will guarantee the achievement of activity goals under possible uncertainty. This is connected with a fundamental feature of new technologies: the development and practical use of a new technology are separated in time. Technology development corresponds to the initial stages of its life cycle. Prediction can be very difficult and treated as an independent complex activity. This process has specifics but, for predictive purposes, the impact of uncertainty can and must be structured into the two general-system attributes considered above. First, these are the primary sources of uncertainty [1] as follows:

- the uncertainty in the environment—the external demand and external conditions, requirements and norms;
- the uncertainty in the technology and subject matter—the means, methods and factors;
- the uncertainty in the actor—recognizing the needs, goal-setting, executing the actions, evaluating the result and decision-making (acting as the subject of CA or not).

Second, these are the structuring binary attributes [1] as follows:

- Will the result have the planned properties or functions?
- Will the final properties and functions of the result yield the desired effect in the interaction with the environment?
- Will the properties of the result match the future demand when the result is presented to the customer?
- Will the environment match the current prediction of its state when the result is presented to the customer?

On the one hand, a technology is a collection of interconnected elements i–v; on the other, any SEA can be described by the structural, cause-effect (causal) and process models [1]. Due to the fractality of CA elements, a complete description must include the models of all lower elements in the logical structure of CA. Such a collection of models reflects the conditions, forms, methods, means and criteria of goal achievement, i.e., the technology elements. Therefore, at the general-system level we may claim the following [1]: the technology of any CA element is described by the structural, cause-effect and process models together with the models of lower SEAs and elementary operations.

Technology testing. Any activity is reflexive in the following sense: generally, a newly created technology and its result can and must be verified and/or falsified (at the epistemological level) and/or checked at the technology implementation level. Such a check can be called *testing*.

On the one hand, the technology of CA has a considerable share of specific components (the technologies of elementary operations). On the other hand, it includes the logical, cause-effect and process models of CA, which determine its general-system components.

Therefore, technology testing (including the completeness of its description) also has general-system specifics. They will be formulated as a list of testable conditions that must be satisfied for a new technology.

I. The structural completeness and consistency of technological goals—the logical structure and the subgoals structure of the central SEA.

II. The availability of a specified actor and technology (SEA or elementary operation) for each technological goal.

III. The availability of specified characteristics for the subject matter of CA for each of the subgoals, which can be used to check their achievement and evaluate the efficiency.

IV. The mutual logical consistency of the actors and technologies associated with the subgoals structure and also with the resource pools for their support and organization.

V. The availability of a specified resource allocation mechanism for the functions performed by the actors of all lower SEAs or for the technologies of all lower elementary operations (in particular, for making the goals and preferences of individuals as the actors of SEAs consistent with the goals of the central SEA).

VI. The availability of specified uncertain events and response rules for them within the central SEA (procedures for decision-making or escalating current problems to higher SEAs).

VII. The consistency and coordination of the logical and cause-effect structures.

Tests I–VII can be implemented together within a separate reflexive operation or independently embedded into separate operations.

Life cycle of technology. As was noted in [1], a special case of resources is knowledge and technologies (technological knowledge as operational knowledge; see Fig. 1.2).

The life cycle of knowledge as an object is simple and includes three phases as follows:

1. development/creation (an analog of the organization and design phase);
2. productive use, with a possible return to the first phase (an analog of the technological phase, also called the implementation phase);
3. an infinite existence in form of historical data, with a possible return to the second phase (the reflexion phase).

If knowledge is used as an element of some technology (operational knowledge), then the life cycle of its pool (Fig. 1.2) is similar to the life cycle of a resource pool [1].

Technologies and hierarchical structure of CA. From the viewpoint of the logical structure of CA, the relations of upper and lower SEAs in terms of the subject matters and actors of their activity can be described by Table 1.2.

Here the columns correspond to the upper (superior) SEAs while the rows to the lower (subordinate) ones. Each cell reflects the relation between upper and lower SEAs. The subject matter of lower SEAs can be any element of the upper SEA. In other words, lower SEAs may be organized for creating/executing/transforming a subject matter/result, technology or actor (OTS).

As is well illustrated by Table 1.2, knowledge—technology description—represents a connecting link between the animate (staff/OTS) and inanimate (product/technological complex) elements and subject matters of activity.

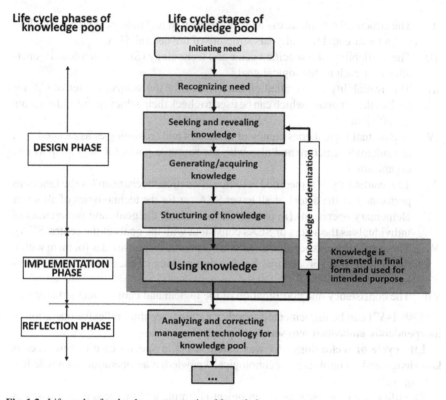

Fig. 1.2 Life cycle of technology as operational knowledge

Table 1.2 Relations between subject matters and actors of SEAs

		Upper SEAs		
		Material product	Knowledge	OTS
Lower SEAs	Material product	Creating product components	Creating products (equipment) for technology	NO (indirectly through knowledge and technologies)
	Knowledge	Creating technological process for product manufacture	Creating components of technological process	Creating technology of OTS functioning
	OTS	NO (indirectly through knowledge and technologies)	Staff training for executing technological process	Transforming OTS elements

Complex activity has a logical structure: the goal of each SEA is decomposed into the lower subgoals, which have cause-effect relations. The technology of CA determines the cause-effect relations between the goals of CA—the SEAs and elementary operations. For example, design, erection, maintenance and other works must be performed in a definite sequence for obtaining a required result.

Creative activity and technologies. Creative CA and creative SEAs are an activity that generates an a priori uncertain demand for the results of an a priori unknown activity whose technology has to be designed in the course of the former activity. In other words, the need for technology design is a criterion of activity's creativeness.

Creative activity (creative SEAs) is an activity with an incompletely defined (partially known) technology at the time of its beginning. Therefore, the technology of creative activity is designed during its implementation. This technology is unknown due to the uncertain demand and/or a priori uncertain specification of its result. Creative activity produces a result that is not completely specified at the time of its beginning. For example, consider the activity of design and production managers, researchers, the producers of movies and shows, the partners of law firms, etc. The actors of creative CA independently determine the structure and characteristics of complex result and hence the structure and technology of activity. In fact, they are the engineers of activity (as a system) and also the engineers of its result (as a system). A fundamental difference between the creative and replicative (and regular) CAs is that the former's structure contains at least one fragment in which the subject matter of activity is the technology of another (lower) fragment of CA. This follows from the need for a new technology to be designed during activity implementation.

Design of optimal technology. To a considerable degree, technology design is a *specific* process. Therefore, it can be optimized only up to the development of several alternatives, with further choice of a best one. (Note that the number of such alternatives affects the computational complexity of this process.) In many cases, technology design is heuristic, which requires the use of appropriate error estimation methods. Both aspects (complexity and errors) will be considered in Chap. 4 of the book.

At the same time, technology design establishes some requirements to resources, which leads to several optimization problems as follows:

– determining an optimal set of resource pools in accordance with the needs of technologies of different CA elements;
– determining an optimal amount of each resource pool;
– maintaining the characteristics of resources within given ranges in an optimal way during their life cycle.

The design of new technologies includes (a) *a general-system part* (here, of secondary role) with the above-mentioned resource optimization problems and (b) a specific part with the design of technologies of all elementary operations and also the logical and cause-effect structures. Due to these specifics the corresponding optimization problems cannot be formulated at the general-system level. However, for this part a series of general-system recommendations or requirements can be formulated, which will serve as bases for developing alternatives (and choosing a best one among them).

First, all technologies can be divided into groups by their levels of maturity, testedness, suitability, or readiness. Different tools of organization and management have to be used for the elements of CA whose technologies belong to different groups. In the most general case, there exist three groups of technologies as follows (in a certain sense, groups 1 and 3 are polar).

Group 1. Well-known technologies used without any modifications or with slight modifications not affecting the technological uncertainty. For example, the replication of a well-known component (assembly or end product) using a well-known technology on a new production site; the production of a well-known component using a well-known technology with a small change of its dimensions. In these situations, the technology of a new element of activity is a priori known and is actually copied.

Group 3. Fundamentally new technologies with a considerable level of the technological uncertainty for which the possibility of successful implementation can be evaluated by subjective expertise only. For example, the use of new processing principles for raw materials in manufacturing; repair and maintenance works in uncertain conditions (the significant characteristics of an object that affect the repair and maintenance man-hours and also the result of these works are a priori unknown and can be evaluated only in their course). In these situations, the technology of activity is designed during the activity itself.

Group 2. The technologies not included in groups 1 and 3 cover all intermediate situations with a considerable a priori known part of technology. *Technological uncertainty* can be measurable or true, but without any considerable effect on the final result.

Second, introduce the general-system bases that will be used in further recommendations. Recall that some approaches to eliminate uncertainty were considered in the book [1]. One of the important conclusions established there is as follows. For the case of a true uncertainty with completely unknown characteristics of events, all or some components of the activity and/or environment turn out to be insufficiently studied. Therefore, in this case, activity represents the acquisition of new knowledge, causing modifications in its technology during implementation.

Hence, a natural base for further recommendations is a more precise and detailed fixation of elements of activity with small level of uncertainty. This allows us to specify the domain of uncertainty, the existing knowledge, and to get focused on the really uncertainty elements. Consequently, the general-system recommendations have to deal with a more precise and detailed description of all elements of CA in terms of subject matter, actions, technologies and other bases as well as with the localization of uncertainty or the extension of "the domain of certainty."

Thus, the following recommendations on the design of specific technologies can be suggested.

(a) Structure the subject matter of CA, with sufficient level of detail, for a more pre-
    cise localization of its highly uncertain elements and, at the same time, identify
    its standard (typical) elements for which ready-made solutions can be adopted.
    For example, a key trend of modern production technologies is a wide use of
    purchased components for end products (about 65% cost of a modern car is
    accounted for standard parts).
(b) With sufficient level of detail, structure the technologies of CA. This recommen-
    dation has a close connection with recommendation (a): separate the elements
    of activity that belong to different groups 1–3 and apply different approaches
    to them in order to improve the efficiency of activity. In practice, this recom-
    mendation covers the detailed specification of technological operations and also
    planning.
(c) Develop and study alternatives, for the technology of activity and also for its
    subject matter and elements. The development of several alternatives under high
    uncertainty is equivalent to the formation and testing of several hypotheses about
    an a priori unknown object, which increases the amount of knowledge about it.
(d) Employ the scenario approach for prediction. Different scenarios of possible
    events development also increase the amount of knowledge (although, subjec-
    tive) about the uncertain future. This knowledge can be used for technology
    design. For example, the development strategy of a retail bank (a special case
    of CA technology) is designed using some long-term forecasts—scenarios—
    for the dynamics of financial and stock exchange markets, even despite the
    subjectivism of such an approach.
(e) Stipulate for the intermediate and preliminary actions on integrating the ele-
    ments of the subject matter and technology into a unified system. The uncertainty
    in any complex system is directly connected with complexity and emergence,
    which often causes considerable problems during system integration. For exam-
    ple, the design of a new model of a modern aircraft or car includes the design of
    very many heterogeneous systems, assemblies and units. Due to their different
    character, the components of a product are designed by separate groups of engi-
    neers, often from different firms (or even countries). In this case, the integration
    of an end product becomes a complex process consisting of several intermediate
    stages.
(f) Stipulate for the additional actions on refining the needs, i.e., making a more
    precise specification of the desired characteristics of the subject matter. The
    complexity, emergence and uncertainty of the subject matter and goal of CA as
    a complex system induce incomplete knowledge on the needs at the early stages
    of CA. In other words, a natural situation is when the actor and also all external
    users of the CA result (customers) have inaccurate estimations of its goals—the
    target values for the characteristics of the CA subject matter. Therefore, in the
    recent decades the traditional customer feedback methods of needs refinement
    have been supplemented with special management tools used in uncertain needs
    conditions, such as Agile and SCRUM.
(g) Stipulate for intermediate checks of activity—the evolution of the subject mat-
    ter's characteristics—for an early identification of any their deviations from the

required dynamics. In the modern project and program management, product life cycles management, a common principle is to plan and perform intermediate tests called the checkpoints, milestones or gates of decision-making. Each test is a priori specified by several groups of rules. First, the matter concerns the rules determining the times of tests. They can be associated with the execution times of separate tasks (e.g., 10 days after the beginning of works), stages or even steps of activity; an alternative approach is to determine the times of tests depending on the evolution of the subject matter's characteristics. The second group of rules specifies a set of the subject matter's characteristics and their combinations to be tested. The third group describes all necessary actions to be performed after each test depending on its results.

(h) Stipulate for a checking procedure for the CA technology rather than for its result. Such a procedure can be used (1) to identify incipient problems as soon as possible and (2) to reveal the manifestations of problems—any deviations of the subject matter's characteristics from the required dynamics—and also their causes dictated by the technology. In the recent decades, this approach has become widespread in practical activity: different quality standards (e.g., ISO9000) and production maturity models (e.g., CMMI) are being intensively used now.

(i) Stipulate for a checking procedure for the implementation of activity by lower actors. This recommendation is connected with recommendation (h), also guaranteeing an early identification of problems before the CA results of lower actors are submitted to customers. This procedure involves the checks of intermediate results and, in the first place, the CA technologies of lower actors. In practice, large companies and government authorities require that their suppliers (lower actors) undergo an independent quality certification.

The above recommendations lead to more precise estimations of the complexity and resources required for the elements of activity, which is achieved by decomposing the activity and its subject matter into detailed elements. As a result, the resource intensity of the activity can be reduced using the differentiated optimization of all necessary resources of the activity elements with different level of uncertainty, including the augmentation of regular activity elements—the tested technologies and also the verified results of activity. Additional tests and intermediate integration can be used to identify problems at earlier stages (thereby avoiding potential losses due to unreasonable actions) and to choose an alternative path of activity implementation.

Of course, any of these recommendations imposes extra cost to maintain additional elements of organization and management. Therefore, they should be followed mostly during the design of fundamentally new technologies (group 3) rather than during the use of well-known technologies (group 1).

Management of CA technology. Two important types of activity—management and organization—were considered in detail in the book [1]. The components of organization (analysis, synthesis and concretization) as well as the components of management (organization, regulation and reflection) were described. As was demonstrated there, the subject matter of organization and management for CA is the aggregate of complex activity itself and the actor implementing the latter (OTS).

In the book [1], the general management problems of OTSs were studied in the context of coordinating the interconnected life cycles of the corresponding structural elements of activity. It was shown that an *OTS is managed* through a coordinated control of all interconnected life cycles of the structural elements of the CA implemented by this OTS. The tools for solving the OTS management problem were defined as the following components of management: *synthesis* (technology components management based on information models; resources pools management) and *concretization* (network planning and scheduling, resource allocation; interests coordination for different actors). It was discovered that the OTS management problem has to be solved by eliminating *measurable uncertainties* (more specifically, by considering different response scenarios for them) and also by performing multiple iterations of the sequential solution procedure if necessary due to a possible occurrence of the true-uncertainty events during the life cycles of all CA elements.

Managing the technology of CA is a complex activity towards creating the information models of its components (including resources) and maintaining them in actual state through modernizations.

1.2 Technological Adaptivity, Cyclicity and Regularity of Activity of Modern Enterprises

The modern stage of global economy development is remarkable for several trends determining some peculiarities of the firms, governmental agencies and other actors of the international economic system, in particular, their activity. Consider these trends and also the associated peculiarities of the modern enterprises and their complex activity.

Social digitization and the development of new approaches to organize and manage the economy (networked, extended and virtual enterprises, global and service-based production, the Internet of Things, to name a few; for example, the new technologies of management were surveyed in [5]) have resulted in deep integration of different enterprises and their activities. Really, it is not enough to connect different sensors, executing devices and controllers of machining centers or automatic warehouse complexes, to integrate automatic control systems with computer-aided manufacturing systems at the level of shop floors or the entire enterprise. Integration has become global, and presently a common form of organization is the so-called *extended enterprises*, i.e., the sets of enterprises and firms united by the same technological processes and relations without legal or financial integration. For the extended enterprises, a major role is played by their technological relations rather than by their pattern of ownership, organizational structure or stock capital.

In management, the ideas of changing organizational paradigms, replacing the rigid managerial organizational structures with the platforms and functional houses [6] (the pools of homogeneous resources) and also increasing the flexibility and speed of response to the varying environment are becoming more and more popular. This means that managerial relations are gradually transferred from the rigid organizational structures to the flexible *technological relations*.

An example of extended enterprises is Boeing Civil Aviation, a company that manufactures the Boeing 787 Dreamliner. About 60% components of this commercial aircraft are supplied by nearly 20,000 subcontractors around the world (Japan, Italy...), which are operating in an integrated technological chain determined by the parent company.

In different fields of practical activity, e.g., manufacturing and control of organizational and technical systems (firms, organizations, projects), the concept of life cycles (LCs) has become widespread recently. Following the definition given in [7], a life cycle will be understood as the evolution process of a system, product, service or another object, from its origin (or design concept) to utilization (or disappearance).

The life cycle is often treated as a set of *stages* (perhaps, parallel or overlapping with each other in time). In [7], the general LC stages of a complex artificial system were identified as follows: concept, design, production/development, application, maintenance and utilization. The concept of LC is widely used for organizations, businesses, project programs, employees, production assets, technologies and knowledge.

Natural cycles are connected, e.g., with the multiple repetition of

- a typical operation;
- a production process for a component or a service procedure;
- a working shift or working day;
- an accounting or schedule period.

These examples point to the existence of CA cycles and their heterogeneous lengths. Some cycles are parts of the other, thereby forming complex hierarchies. Consider a general diagram of CA cyclicity presented in Fig. 1.3.

Another relevant trend [5] consists in announcing a series of national programs (e.g., Industry 4.0 in Germany and Smart Manufacturing in the USA) towards the development of industrial enterprises that will be capable to combine

- the high efficiency of large-scale productions with the high personalization of job-order or even artisan productions;
- the fast and efficient design or modernization of technologies with the regular and standard computer-aided, digitalized and robotized productions.

Also note the fast varying political, economic, social and technological environments of the enterprises as well as the dynamics of the demands for their products and services.

Therefore, today the global economy represents a set of different-scale enterprises, which come into existence, implement the life cycles of their complex activities and disappear; have complex relations with each other; form new complex structures

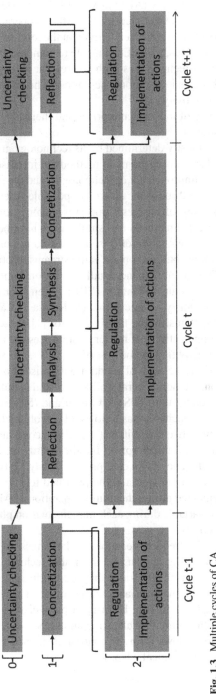

Fig. 1.3 Multiple cycles of CA

and become their elements. Hereinafter, such enterprises will be described by the term *adaptive extended enterprises* (AEEs). An AEE is an extended enterprise (a special case of organizational and technical systems) that is operating under dynamic uncertainty and has high adaptivity of their activity at the level of technology design and modification (In other words, an AEE is capable to combine the fast and efficient design or modernization of technologies with the regular and standard computer-aided, digitalized and robotized operational activities). This property of AEEs will be called technological adaptivity.

Increasing the efficiency of modern enterprises and their activity is an issue that deserves special consideration.

The efficiency of CA is mostly determined by its technology and also depends on the realized uncertainty [1]. Hence, the main methods to increase the efficiency of modern enterprises are the improvement of technologies and the reduction of uncertainty. They are implemented through the maximum possible ordering and regulation of activity. The reengineering of business processes, quality control based on the ISO9000 standards, the LEAN methods and similar ones have become widespread here. The improvement and regulation efforts are focused not only on the process flowcharts of key productions but also on the value streams of other lines of activities such as logistics, infrastructure, finance, staff, etc. A conventional approach is to use information enterprise resource planning (ERP) systems for detailed planning. As a matter of fact, the majority of enterprises (in Russia and all over the world) are using different ERP systems. In the terminology of [1], these trends mean that most elements of the routine activity of modern enterprises are regular, i.e., a total "regularization" of AEEs activities can be observed.

A common feature of the activity of modern enterprises concerns continuous improvements. In this context, note several well-known and popular approaches like TQM, the Toyota Production System (TPS), Six Sigma and 7S Framework McKinsey. The main ideas of these approaches are as follows: (1) involving the entire staff into the improvement of all enterprise's activities; (2) perform continuous improvements during the routine operation of an enterprise. In other words, at the reflexive phase of the CA life cycle, the efficiency of an enterprise is deeply analyzed; at the design phase, the CA technology is improved and the modified technology is used at the next cycle. This is implemented during the routine operation of AEEs. Interestingly, the technology remains invariable during each implementation phase, which allows performing the regular activity.

Thus, AEEs combine the frequent technological changes/improvements at the design phase with the use of the regular and invariable technologies at the implementation phase. Figure 1.3 well illustrates this peculiarity, also emphasizing the simultaneous character of the managerial activities of the actor: the technological changes/improvements (level 1 in Fig. 1.3) run in parallel with CA implementation (level 2), together with uncertainty checking and response generation (level 0). Recall that this peculiarity follows from the continuous improvement of efficiency in combination with the technological adaptivity of AEEs.

1.3 Model of Complex Activity of Adaptive Extended Enterprises

Now, introduce a formal description for the life cycle of the activity of an adaptive extended enterprise with the above-mentioned peculiarities. Further considerations will be based on the process model of an SEA—the life cycle model of CA suggested in the book [1]. This model includes three phases of CA, namely, design, implementation and reflection. Also the detailed structure of the managerial activities of the actor during the entire life cycle of CA will be used, as is illustrated in Fig. 1.4.

Consider the life cycle model of the CA of an AEE—the process model—in the BPMN format[3] [8]; see Fig. 1.5.

The design phase is intended to develop a technology and an AEE itself. At this stage, the actor's activities [1] are the components of organization as follows:

- *analysis* (1—establishing demand, analyzing capabilities, external conditions and previous activity);
- *synthesis* (2—managing technology components based on information models: creating the logical, cause-effect and process models as well as the technologies of lower elements and maintaining them in actual state through modernizations; 3—managing resources pools: assigning and modernizing resources);
- *concretization* (4—network planning and scheduling, allocating resources; 5—coordinating the interests of different actors).

A newly created technology is used for CA at *the implementation phase*. The actor performs regulation (6) and also executes the actions (7). The life cycle of CA ends with *the reflection phase*, at which the results of CA are evaluated (8).

Recall that the activity of an AEE is cyclic, and this property implies the following. Once establishing the demand in the course of analysis (1) and creating the technology components (2), during the implementation phase the actor repeatedly executes the activity in some cycle, which will be called *productive*. (Accordingly, the activities

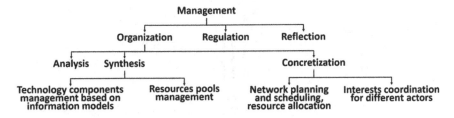

Fig. 1.4 Detailed structure of management components

---

[3] The BPMN format uses the following notations: rounded rectangles as operations or actions; arrows as control flows—the sequences of transitions between actions; circles as different events (thin boundary—initial event; thick boundary—terminal event; double boundary—event of uncertainty occurring during action implementation); diamonds as control points—branching and merging of control flows, including parallel execution and conditions checking.

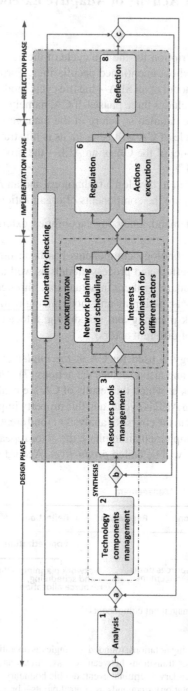

Fig. 1.5 Life cycle model of CA of AEE

that represent this cycle will be called productive activities; e.g., see activities 3–8 inside the grey rectangle in Fig. 1.5).

Recall that an AEE is technologically adaptive, and this property implies the following. The CA technology of an AEE undergoes periodic modifications or modernization (2); clearly, modernization (2) is preceded by the evaluation of the past activity (8).

Recall that an AEE is operating continuously, performing CA in cycles, and this property implies the following. Uncertainty checking, reflection and technology modernization are performed in parallel to CA implementation, as is shown in Fig. 1.3. That is, during a definite productive cycle the CA performed at the previous cycle is reflected and its technology is modernized. The improved technology will be used at the next productive cycle.

Recall that the activity of an AEE is regular, and this property implies the following. The technology of CA elements remains invariable during the productive cycle in which they are implemented; their technology is modified or modernized only during one of the preceding cycles (the technologies of the productive activities 3–8 are fixed during their execution).

Recall that the CA of an AEE is uncertain, and this property implies the following. The events of measurable and true uncertainty may occur during the implementation of life cycles. The possible occurrence of measurable uncertainty must be considered during technology components design; hence, such events do not violate the implementation of the life cycle of CA, which corresponds to the repeated execution of the productive activities (3–8 between points b and c). On the contrary, the events of true uncertainty generate conditions under which the technology becomes inadequate. As a result, the technology must be modernized; see the transition to point a and the execution of activity (2)—technology components management—in Fig. 1.5.

1.4   Management of Technology Components Based on Information Models

The list of all technology components of complex activity follows from the system of descriptive models suggested in [1]. Actually, the technology components of each SEA [1] are:

- the logical, cause-effect (causal) and process models;
- the technologies of lower elements[4];
- the technologies of SEAs;
- the technologies of elementary operations.

---

[4]The particular cases of the lower CA elements are managerial activities—resources pools management, network planning and scheduling, interests coordination for different actors, regulation and evaluation (reflexion).

Because the technology of an SEA includes the technologies of all lower SEAs in accordance with the logical structure, the technology is fractal, like the complex activity itself. Hence, the management of technology components also has fractal organization: this activity is implemented through iterative self-addressing (Fig. 1.6).

The technological structure of an SEA consists of the logical, cause-effect and process models and also the relations and design sequences of the technologies of all its lower elements. Accordingly, managing the technological structure means creating and maintaining this structure in an adequate state depending on the environment.

The elements of "Managing technology components" are decomposed into the lower-level elements and form a fractal hierarchy. The elements of the "Managing technological structure" and "Managing technology of elementary operations" blocks depend on the specifics of a given subject matter and also on the peculiarities of a given activity. They do not need further specification (a more detailed description) at the system level. The technologies of elementary operations and the technological structure can be therefore united and called *the elementary technology components* (ETCs). Following this approach, the other technology components will be called *the complex technology components* (CTCs).

Consider a management model for the ETCs starting from their creation. The creation of the ETCs is **heuristic**; generally speaking, this process cannot be described in detail or formalized (of course, if the matter concerns the creation of a new technology rather than the redesign of a partially or completely known technology). Therefore, it can be represented by a single element—an elementary operation (the structuring of heuristics makes no sense). However, the life cycle of any activity includes the reflection phase and the creation of the ETCs is not an exception here: the adequacy of each ETC has to be evaluated for different states of the environment. Also note that the heuristic is preceded by the actor's decision to perform activity.

Thus, the life cycle of designing the technology of the ETCs as an activity can be described by the process model in Fig. 1.7. It includes three main phases as follows:

1. the analysis phase (the actor decides to implement activity);
2. the heuristic phase (the actor generates a heuristic—a draft ETC—also called an alternative);

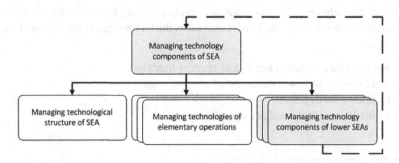

Fig. 1.6 Management of technology components: logical structure of activity

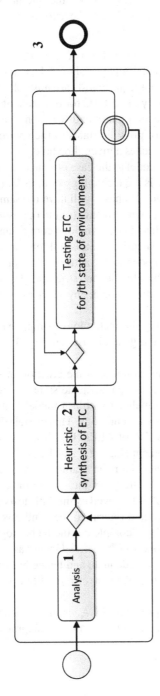

Fig. 1.7 Process model of managing ETCs

3. the reflection phase (the actor tests whether the alternative complies with the goal–requirements–demand chain. If not, the actor returns to phase 2, repeating it until an adequate alternative is obtained).

Adequacy tests can be performed in the form of mental experiments; model experiments with mathematical, computer or physical models; natural experiments. In the general case, we may assume that adequacy tests are multiple repeated and each test is used for evaluating the adequacy of an ETC for one state of the environment that occurs during this test. The reflection phase ends as soon as a maturity measure of tests (e.g., the share or number of possible states of the environment for which the testing procedure is complete) reaches a given threshold.

Now, generalize this process model to the case of managing the complex technology components. Generally, this is the technology of an SEA (Fig. 1.8).

The analysis (1) and reflection (3) phases will have the same form as in the case of managing the ETCs, as there are no grounds for the opposite.

As indicated by the complete list of all technology components (see the analysis in the beginning of this section), the heuristic phase will represent a certain sequential-parallel combination[5] (2) whose nodes are the activity elements on managing an ETC (a) or the technology of an SEA (b). The staff and structure of this combination depend on the specifics of the subject matter. Hence, any other rules cannot be established or used at the general-system level.

Thus, the activity on managing the technology components is reflected by *the process model*; see Fig. 1.8 for the general case and Fig. 1.7 for a special case of ETCs.

This managerial activity includes not only the creation of the technology components but also their maintenance in an adequate state during use. Therefore, the management problem can be formalized within a model that well describes both processes—the creation and use of a technology. This model (Fig. 1.9) actually integrates the life cycle model of the CA of AEEs (Fig. 1.5) with the process model of technology components management (Figs. 1.7 and 1.8).

The integrated model illustrates the main feature of the technology component management process—multiple cyclicity. In the course of technology creation or modernization, direct synthesis (2) is followed by multiple tests (3) of the synthesized technology. If a test fails, the process returns to the synthesis stage (from d to a). As soon as all tests are successfully completed, the technology is repeatedly used within the productive cycle $(4, b \rightarrow c \rightarrow b \rightarrow \ldots)$. Under varying environment (e.g., demand), the technology becomes inadequate and hence has to be modernized; see return from c to a. All these cycles together form the life cycle of a corresponding element of complex activity.

As a result, the management problem of CA technology consists in the management of the multiple interconnected cycles within the integrated model in Fig. 1.9.

---

[5]The staff and relations of nodes in structure (2) are somewhat conditional.

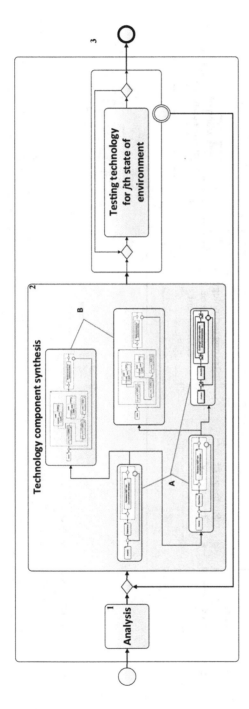

Fig. 1.8 Process model of managing technology components

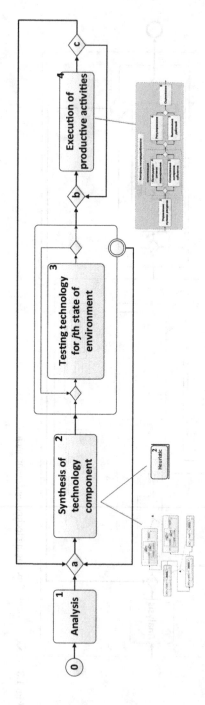

Fig. 1.9 Integrated model of technology component management

1.5 An Overview of Well-Known Models and Methods

Formally speaking, the management process of technology components is the sequential repetition of cycles within the integrated model (Fig. 1.9); see the details in the previous subsections of this chapter. Such problems arise in many fields of knowledge (e.g., complex systems testing, software analysis and testing), and each field suggests specific solutions. Consider a series of well-known models and methods for solving them:

- complex systems analysis—testing and verification of characteristics [9];
- software testing [10–16];
- knowledge management and elicitation/acquisition [17–19];
- large-scale manufacturing and its efficiency improvement during adoption [20–22];
- learning in pedagogics, psychology, human and zoophysiology [23–26] and machine learning [27];
- knowledge testing for trainees in pedagogics [28].

Like technology design, all these problems have uncertainty and are often described using probabilistic models and/or the framework of random processes.

The testing procedure of complex systems (in particular, aircraft complexes) is represented as a hierarchical structure in which nodes describe the tests of elements, units and systems. The efficiency of product's components is assumed to have an exponential (or logistic) dependence on the duration of tests; as a rule, the rate of growth of this efficiency is assumed to be proportional to the unreliability detected at a current time. The expected test times required for reaching a given efficiency level and also the corresponding cost are calculated at each level of the hierarchy. The total expected test time of a system (product) is the sum of the expected test times at each level of the hierarchy. Actually, the assumption that the efficiency of the system's components depends on the duration of tests is not completely justified while the integration of the components' tests as the sum of their expected test times seems a strong simplification (although this general scheme can be taken as a basis for further development and correction). The optimization problem of the testing procedure is written using the well-known approximations of the random functions in terms of their means and variances. The testing problems of different products and mathematical models are often posed as the problems of hypotheses verification and experiment planning.

Model checking and its modification—statistical model checking—are popular methods to test complex systems and complex software [9, 10]. These methods are used for complex systems with a finite set of states and quantitative properties specified by logical expressions. Such an approach allows measuring the correspondence between system properties and their required values. A stochastic system is tested by verifying the hypothesis that its properties satisfy given requirements. The logical descriptions of the system can be used to integrate elementary tests into complex ones. General testing approaches were presented in many classical works; for example, see the Guide to the Software Engineering Body of Knowledge (SWEBOK), version 3.0, in [13].

Another popular software testing method is regression testing and its numerous variations [11]. Regression testing is intended to verify exploited software and to prove its quality after changes and modernizations. The collections of tests are gradually growing in size following the rapid development of software, which makes the execution of all tests for each change very costly. Regression testing includes such techniques as the minimization, selection and prioritization of tests, which are implemented as formal procedures. Minimizing the collection of tests, a software engineer eliminates all redundant test examples. Selecting test examples, a software engineer activates the tests directly connected with the last changes. Prioritizing test examples, a software engineer adjusts a sequence of tests so that the errors can be detected as fast as possible.

Among other common methods, note model-based testing, different sequential testing procedures and sequential analysis procedures [29], simulation modeling [30], special testing methods for cyber-physical systems [12], etc.

In the paper [30], the coordinated planning problem for a group of autonomous agents (mobile robots) on a pendulum plane was considered. The goal of the group was to stabilize the plane. This problem was solved using an original collective learning procedure with the cyclic generation of learning signals and the statistical reinforcement of "the skills."

Today, there exist many descriptive models for the process of knowledge elicitation/acquisition; all of them are implementing sequential processes to analyze the subject matter and improve the models. Consider several examples as follows. In [19], a general decision support algorithm based on the analysis of large mixed data was suggested. As was claimed in [17, 18], the stochastic and deterministic knowledge are supplementing and improving each other; a stochastic model of acquired knowledge based on the diffusion approximation was also introduced. The paper [17] was dedicated to the optimal Bayesian agent—an algorithm that describes the process of knowledge elicitation in the course of sequential observations of a stochastic environment with a denumerable set of states. The algorithm rested on Solomonoff's theory of inductive inference.

In [23], the generalization methods of knowledge acquired from empirical observations were considered. Knowledge was synthesized by a clustering algorithm using the identification of statistically significant events. The algorithm with a probabilistic information measure performed the grouping of ordered and unordered discrete data in two phases as follows. During cluster initiation, the distribution of the distances between nearest neighbors was analyzed to choose a proper clustering criterion for the samples. During cluster refinement, the clusters were regrouped using the events covering method, which identified the subsets of statistically significant events.

The potential improvement of manufacturing efficiency as the result of technology adoption had been discussed by economists since the early development of machines in the 19th century but the per-unit cost reduction effect for large production outputs was first described by Wright as far back as in 1936; see [22]. Wright's approach postulated the exponential model of the learning curve, also known as Wright's curve in management and economics. Wright's model was generalized by Henderson (Boston Consulting Group) in his paper [21]. During the research performed by Boston Consulting Group in the 1970s, the specific cost reduction effects were identified for different industrial sectors; they varied from 10 to 25% under double output increase.

Recall that *learning* is the process and result of acquiring an individual experience. In pedagogics, psychology, human and zoophysiology, iterative learning models [31] describe the process of learning in which a learned system (can be living, technical or cybernetic) is repeating some actions, trials, attempts, etc. over and over again for achieving a fixed goal under constant external conditions. In [31], tens of the well-known and widespread iterative learning models were surveyed and a general model combining the properties of separate models was formulated. Both the separate and general learning models have restrictions as follows. First, they postulate certain learning laws. Second, they do not provide for proper integration processes of partial learning elements into a complex learning system.

In pedagogical measurements, the methods of item response theory [28] have become very popular in recent time. This theory is intended for evaluating the latent (unobservable) parameters of respondents and test items using statistical measurement models. In item response theory, the relation between the values of the latent variables and the observable test results is defined as the conditional probability of correct answers to the test items by the respondents. The conditional probability is given by the logistic curve or the Gaussian probability distribution. The most widespread models of this class are the Rush and Birnbaum models, which use the specific values of the coefficients of the logistic curve.

Summarizing this short overview of the well-known results, we emphasize that the models mentioned are actually some elements of a cycle of the integrated model in relatively simple form: (1) without the iterativeness and fractality of such cycles; (2) with the postulated character of the basic laws (e.g., the exponential or logistic relation between the efficiency of the product's components and the duration of tests; the exponential or logistic relation between the learning level and time). Generally speaking, the basic laws are a consequence of more sophisticated processes, which have to be analyzed and modeled. At the same time, the results of the book [1] have been adopted to study the peculiarities of this cycle and reflect them by the general-system integrated model (Fig. 1.9). Consequently, this model further generalizes, specifies and refines the well-known models.

Using the framework of the integrated model (Fig. 1.9), we will formulate some optimization problems for the management process of CA technology components.

1.6 Management Problems for Technology Components

In the life cycle of complex activity (Fig. 1.9), the design and use of its technology is a process that includes the creation of elementary technology components (ETCs) and also their integration into the complex technology of CA. First of all, management problems have to be solved for the ETCs; then, based on these solutions, the ETCs have to be integrated into the complex technology components (CTCs), the entire complex technology, and its management.

Management problems for elementary technology components. The integrated model (Fig. 1.9) describes a cyclic repetition of tests for the ETCs, during the preliminary testing (block 3 in Fig. 1.9) and also during the life cycle implementation (block 4 in Fig. 1.9). Hence, a natural approach is to treat the creation of the ETCs as a discrete process, assuming that at each time (one preliminary test or one life cycle) the environment takes precisely one state from *the set of possible states of the environment* (SPSE). For solving this problem, suppose the SPSE can be partitioned into a certain number of non-intersecting subsets so that all states of the environment belonging to the same subset are indistinguishable. Therefore, let the SPSE be finite and also let the environment take precisely one state from the SPSE at each time.

If at some time the environment evolves to a new state never observed before, then an event of *uncertainty* occurs. This event leads to additional cost for creating or adapting the ETCs to the new conditions. If the environment returns in this state at one of the subsequent times, then no additional cost is required.

Then the design process of the ETCs is completely characterized by the dynamics of the states of the environment: which values from the SPSE the environment has already taken (and how many times) and which has not. Introduce the maturity level of a technology, which describes its preparedness for use. (This is an analog of the learning level; the sequence of values of the learning level is called the learning curve). Define the maturity level of a technology as the share of the states of the environment for which the technology has been tested or adapted during the past times, or the probability that at the next time the environment will take one of its previous states.

A natural formulation of the management problem is to reach a required learning level (a required maturity level of the technology) as fast as possible or using as few resources as possible. Thus, the optimized and/or restricting parameters can be the consumption of resources or time.

In the general case, the technology management process can be optimized by:

- partitioning all states of the environment into non-intersecting subsets so that the states from each subset are equivalent;
- choosing a sequence of the exhaustive search of all states of the environment for testing;
- redistributing these tests between the design level of the ETCs (block 2 in Fig. 1.9) and the level of the SEAs after the integration of the ETCs (block 3);
- allocating limited resources among separate ETCs, with or without considering the ETC design process (its beginning and duration);
- determining an admissible amount of attracted resources (in terms of risk) for supporting all ETCs.

The heuristic operations of technology design (block 2 in Fig. 1.9) and technology testing (block 3) are specifics and hence cannot be managed or optimized at the general-system level. Therefore, the heuristic operations will be assumed to have known execution times and resource consumptions (in a special case, random variables with given distributions). In addition, these operations will be assumed independent of each other so that their connections are implemented through given cause-effect (causal) relations between the design processes of separate technology elements; the list of the CA elements and the logical and cause-effect models will be assumed fixed.

The actor's knowledge about the environment and his/her capabilities to influence the latter can be characterized by different factors (see Table 1.3) depending on the CA specifics.

If the actor knows the list of all possible states of the environment and can choose a next state (row 1 in Table 1.3), the design problem becomes trivial: the actor simply rearranges the states in the descending order of their probabilities, sequentially choosing them until a required maturity (learning) level is reached. This algorithm will minimize the time and cost of reaching the required learning level.

The setup in which the actor does not know the states of the environment but can choose them at his/her own will (row 2 in Table 1.3) is inconsistent.

The other setups are consistent and lead to Problems 1–4; see Table 1.3. In all of them, a next state of the environment is uncertain and does not depend on the actor.

If the environment uncertainty is true (i.e., the actor has no grounds to describe the environment by some laws and/or restrictions), the problem becomes degenerate.

In the case of measurable uncertainty, the actor can use stochastic, fuzzy, interval or some other models of the environment to eliminate the uncertainty.

Problem 1 (the actor knows the list and probabilities of all possible states of the environment) surely arises and is basic in the following sense: the actor performs an initial synthesis of the CA technology using a list of all states of the environment defined by the demand for the CA result; the testing process (block 3 in Figs. 1.7, 1.8 and 1.9) corresponds to Problem 1. The basic problem can be formulated as follows: calculate the maturity level curve of the technology and then optimize in terms of time and available resources.

In a similar way, Problem 4 arises during a continuous repeated use of the technology due to the natural variability of the environment. This setup corresponds to the implementation of the productive activities (block 4 in Fig. 1.9). A reasonable formulation of Problem 4 is to optimize the decision procedure on the identification

Table 1.3 Possible setups of technology design problem

Actor knows the list of all possible states of environment and their probabilities	Actor can manage the choice of environment states	The list of all possible states of environment and their probabilities	
		Are fixed	May vary
Yes	Yes	Trivial problem	
No		Setup is inconsistent	
Yes	No	Problem 1	Problem 3
No		Problem 2	Problem 4

of any changes in the environment's behavior and the adaptation/modernization of a technology component if necessary.

Problems 2 and 3 have intermediate character. They arise (or not) depending on the specifics of the subject matter. Formally speaking, both problems are variations of Problem 4.

Management problems for complex technology components. Complex technology components (CTCs) are formed by integrating elementary technology components (ETCs). Therefore, the integration process of the ETCs into the CTTs. This process is implemented in accordance with the logical and cause-effect structures of CA [1].

The logical structure of CA [1] is defined as a finite acyclic graph that describes (a) the goals structure of all CA elements and (b) the fact that each SEA (and the entire CA, as a special case) is decomposed into a finite number of lower SEAs and elementary operations. In addition to the goals structure, the logical structure also represents a "managerial" hierarchy for the subordinance and responsibility of the actors of all SEAs for their results (the achievement of goals). The logical structure of each separate SEA has a single level, actually reflecting the subgoals of all lower CA elements. Hence, at the general-system level the logical models of any SEAs are equivalent: the logical model of a CA element of an AEE has the fan structure, like the logical model of any SEA.

The cause-effect model of an SEA [1] describes the technological relations between the lower CA elements, thereby determining the order of goals achievement for each CA element. This model is a cause-effect structure defined on the goals set. The structure can be described by a directed graph in which the nodes reflect the goals of all CA elements (or the elements themselves) while the arcs their cause-effect relations. The graph has several properties dictated by its interpretation (the nodes are the goals of the CA elements organized by a lower SEA) as follows.

- The graph contains a unique terminal node corresponding to the final goal of all CA elements.
- The graph is connected, since any subgoals that are not necessary for achieving the final goal become unreasonable.
- The graph includes no cycles, since the decomposition of the goals into subgoals implies that any subgoal is achieved once. If a certain subgoal must be achieved a finite number of times, then it has to be reflected as several subgoals. (Note that the decomposition of a goal into an infinite number of subgoals makes no sense.)
- The graph nodes have correct numbering (i.e., there are no arcs coming from a node with a greater number to a node with a smaller number). The correct numbering of the graph nodes reflects the prior beliefs about the cause-effect relations of the results obtained by the lower activity elements.

Such a graph will be called *a binary network*: any element of this network is characterized by its binary result—the goal is achieved or not. If the goal of a CA element is achieved, then the result can be used by other CA elements. If not, the result is inapplicable for other CA elements.

Also a binary network defines the preconditions to implement the CA elements—the logical functions whose arguments are the results of the immediate predecessor nodes. The simplest examples of the precondition functions are logical conjunction (an element is executed only after achieving the results of all its immediate predecessors) and disjunction (an element is executed after achieving the result of at least one of its immediate predecessors).

Due to the properties of a binary network, it is necessary and sufficient to consider the following integration processes for technology components: (a) sequential integration; (b) parallel conjunctive integration; (c) parallel disjunctive integration. For each integration process, the maturity level of a complex technology component has to be optimized in terms of time and resources.

In this chapter, the management problems of the complex activity technology of organizational and technical systems have been described in the form of the algorithmic models and problems of technology components management. This theoretical formalization can be used for developing the mathematical models of the design and adoption of complex technology components of complex organizational and technical systems. Such models will be described in Chap. 2 of the book.

References

1. Belov M, Novikov D (2018) Methodology of complex activity. Lenand, Moscow, 320 pp (in Russian)
2. Novikov D (2013) Theory of control in organizations. Nova Science Publishers, New York, 341 pp
3. Rebovich G, White B (2011) Enterprise systems engineering: advances in the theory and practice. CRC Press, Boca Raton, 459 pp
4. Novikov A, Novikov D (2007) Methodology. Sinteg, Moscow, 668 pp (in Russian)
5. Schwab K (2016) The fourth industrial revolution. World Economic Forum, Geneva, 172 pp
6. De Smet A, Lund S, Schaininger W (2016) Organizing for the future, McKinsey Q, Jan 2016. http://www.mckinsey.com/insights/organization/organizing-for-the-future
7. ISO/IEC/IEEE 15288:2015 systems and software engineering—system life cycle processes
8. Business Process Model and Notation (BPMN), v2.0.2. http://www.omg.org/spec/BPMN/2.0
9. Henriques D, Martins JG, Zuliani P, Platzer A, Clarke EM (2012) Statistical model checking for Markov decision processes. In: International Conference on Quantitative Evaluation of Systems (QEST), London, pp 84–93
10. Aichernig B, Schumi R (2017) Statistical model checking meets property-based testing. In: 2017 IEEE International Conference on Software Testing, Verification and Validation (ICST), Tokyo, pp 390–400
11. Alagoz H, German R (2017) A selection method for black box regression testing with a statistically defined quality level. In: 2017 IEEE International Conference on Software Testing, Verification and Validation (ICST), Tokyo, pp 114–125
12. Ali S, Yue T (2015) U-test: evolving, modelling and testing realistic uncertain behaviours of cyber-physical systems. In: 2015 IEEE 8th International Conference on Software Testing, Verification and Validation (ICST), Graz, pp 1–2
13. Bourque P, Fairley RE (eds) (2014) Guide to the software engineering body of knowledge, Version 3.0. IEEE Computer Society, 2014. www.swebok.org

14. Legay A, Delahaye B, Bensalem S (2010) Statistical model checking: an overview. In: Barringer H et al (eds) Runtime Verification. RV 2010. Lecture notes in computer science, vol 6418. Springer, Berlin, pp 122–135

15. Patrick M, Donnelly R, Gilligan C (2017) A toolkit for testing stochastic simulations against statistical oracles. In: 2017 IEEE International Conference on Software Testing, Verification and Validation (ICST), Tokyo, pp 448–453

16. Yizhen C et al (2017) Effective online software anomaly detection. In: Proceedings of the 26th ACM SIGSOFT International Symposium on Software Testing and Analysis (ISSTA 2017), Santa Barbara, pp 136–146

17. Orseau L, Lattimore T, Hutter M (2013) Universal knowledge-seeking agents for stochastic environments. In: Jain S, Munos R, Stephan F, Zeugmann T (eds) Algorithmic Learning Theory. ALT 2013. Lecture notes in computer science, vol 8139, Springer, Berlin, pp 158–172

18. Stocia G, Stack B (2017) Acquired knowledge as a stochastic process. Surv Math Appl 12:65–70

19. Wong A, Wang Y (2003) Pattern discovery: a data driven approach to decision support. IEEE Trans Syst Man Cybern 33(1):114–124

20. Crawford J (1944) Learning curve, ship curve, ratios, related data. Lockheed Aircraft Corporation, pp 122–128

21. Henderson B (1984) The application and misapplication of the learning curve. J Bus Strategy 4:3–9

22. Wright T (1936) Factors affecting the cost of airplanes. J Aeronaut Sci 3(4):122–128

23. Chui D, Wong A (1986) Synthesizing knowledge: a cluster analysis approach using event covering. IEEE Trans Syst Man Cybern 16(2):251–259

24. Leibowitz N, Baum B, Enden G, Karniel A (2010) The exponential learning equation as a function of successful trials results in sigmoid performance. J Math Psychol 54:338–340

25. Thurstone L (1919) The learning curve equation. Psychol Monogr 26(3):1–51

26. Thurstone L (1930) The learning function. J Gen Psychol 3:469–493

27. Sutton R, Barto A (2016) Reinforcement learning: an introduction. MIT Press, Massachusetts, 455 pp

28. van der Linden WJ, Hambleton RH (1996) Handbook of modern item response theory, Springer, New York, 512 pp

29. Shiryaev A (1973) Statistical sequential analysis: optimal stopping rules. American Mathematical Society, New York, 174 pp

30. Mikami S, Kakazu Y (1993) Extended stochastic reinforcement learning for the acquisition of cooperative motion plans for dynamically constrained agents. In: Proceedings of IEEE Systems Man and Cybernetics Conference (SMC), Le Touquet, vol 4, pp 257–262

31. Novikov D (1998) Laws of iterative learning. Trapeznikov Institute of Control Sciences RAS, Moscow, 98 pp (in Russian)

Chapter 2
Models of Technology Design and Adoption

The management problem of the CA technology of OTSs has been considered and formalized in Chap. 1 (also see [1]). More specifically, the most important peculiarities of the CA of OTSs have been analyzed and also formal models and a mathematical setup of this management problem have been presented.

In this chapter, the design and/or adoption problem of CA technology is formalized as a mathematical model that generalizes some probabilistic models of learning. The properties and characteristics of the model are studied and expressed in analytical form; some integration processes for technology elements are suggested. As is demonstrated below, special cases of the model include the well-known exponential, hyperbolic and logistic learning curves from the classical theory of learning as well as the models of learning-by-doing and collective learning.

This chapter is organized as follows. In Sect. 2.1, using the results of Chap. 1 the general-system problems of CA technology design, adoption, optimization and modernization are examined and the corresponding mathematical problems are formulated. In Sect. 2.2, the properties of the design and adoption process of a technology are analyzed. In Sect. 2.3, some approximations of the learning curve under different probability distributions of all possible states of the environment are obtained. In Sect. 2.4, the expected times of reaching a required learning level are estimated. In Sect. 2.5, the integration models of technology components are described.

2.1 Conceptual Description of Technology Management Problem

Recall that *the technology of CA* has been defined as a system of conditions, criteria, forms, methods and means for sequentially achieving a desired goal. In Chap. 1, *technology management* has been viewed as an activity for creating technology components in the form of corresponding information models, their integration and maintenance in an adequate state depending on the environment during the

© Springer Nature Switzerland AG 2020
M. V. Belov and D. A. Novikov, *Models of Technologies*, Lecture Notes
in Networks and Systems 86, https://doi.org/10.1007/978-3-030-31084-4_2

entire life cycle of CA. This process has been described using an integrated model in the BPMN format [2] (Fig. 1.9). In a practical interpretation, a technology represents the scenarios of the actor's activity in different external conditions (*states of the environment*).

Now, we will refine the concept of technology management by extending the semantics of the integrated model (Fig. 1.9), which reflects different types of CA as follows:

- *the design* of a new CA technology or its *modernization* (blocks 1 and 2 and also the cycle with block 3);
- the *use* of the CA technology (block 4);
- the identification of a need for technology modernization (transition from c to a);
- the testing or *adoption* of the CA technology, which can be objective (a new technology is considered) or subjective (an existing technology is learned by the actor); see the cycle with block 3.

Despite the whole variety of these types of activity, all of them have a common feature—the same subject matter (CA technology). All these types of activity are intended to modify CA technology, its states or relations with the actor. Therefore, all these types of activity consist in the management of CA technology. (In accordance with the definition given in [1, 3], *management* (control) is a complex activity that implements an influence of a control subject on a controlled system (controlled object) for driving the latter's behavior towards achieving the former's goals.

In the general case, the technology management process can be optimized by:

- partitioning all states of the environment into non-intersecting subsets so that the states from each subset are equivalent;
- choosing a sequence of the exhaustive search of all states of the environment for testing;
- redistributing these tests between the design level of a technology component (block 2 in Fig. 1.9) and the integration level (block 3);
- allocating limited resources among separate technology management operations;
- determining an admissible amount of attracted resources (in terms of risk) for supporting all technology components.

Each of the heuristic operations of technology design (blocks 1 and 2 in Fig. 1.9) and technology testing (block 3) depends on the subject matter, which determines its specifics. Hence, they cannot be managed or optimized at the general-system level. Therefore, the heuristic operations will be assumed to have known execution times and resource consumptions (in a special case, random variables with given distributions). In addition, these operations will be assumed independent of each other so that their connections are implemented through given cause-effect (causal) relations between the design processes of separate technology elements; the list of the CA elements and the logical and cause-effect models will be assumed fixed [1].

The execution of different types of CA within the integrated model (Fig. 1.9) will be represented as a discrete process in which each time is associated with the implementation of precisely one CA element (one of the blocks 1–2–3–4 of the model

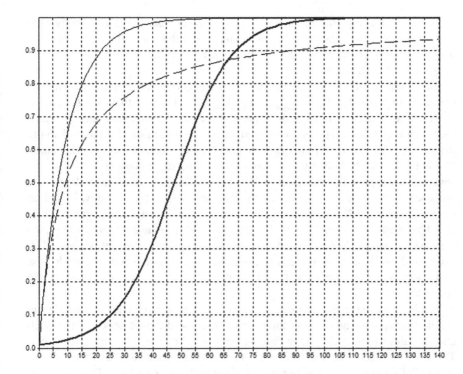

Fig. 2.1 Example of three learning curves: solid line—exponential, thick line—logistic, dashed line—hyperbolic

in Fig. 2.1) under precisely one state of the environment from a finite *set of possible states of the environment* (SPSE).

If at some time the environment evolves to a new state never observed before, then an event of *uncertainty* occurs. This event leads to additional cost for creating or adapting the technology to the new conditions. If the environment returns in this state at one of the subsequent steps, then no additional cost is required. The concept of environment uncertainty implies that the actor is unable to affect the choice of a current state of the environment; the uncertainty will be described using probabilistic methods.

Let the SPSE be composed of K different values. Assume at each time the environment takes precisely one of them regardless of the past states. Denote by p_k the probability that the environment takes the kth value (obviously, $\sum_{k=1}^{K} p_k = 1$).

At time t, the current state of the implementation process of different phases in the technology's life cycle will be described by a K-dimensional row vector $x_t = (x_{1t}, x_{2t}, \ldots, x_{kt}, \ldots, x_{Kt})$ as follows. Each x_{kt} is 0 if the environment has not taken the kth value so far and 1 if it has done so at least once. Within the framework of this model, the kth element of the vector x_t may move from state 0 to 1 but not conversely.

Therefore, the implementation process of different phases in the technology's life cycle is completely characterized by the dynamics of the states of the environment: which values from the SPSE the environment has already taken (and how many times) and which has not.

The maturity level of the technology (its preparedness for use) at time t will be measured by the index $L_t = \sum_{k=1}^{K} x_{kt}p_k$, $0 \leq L_t \leq 1$. (Note that $(1 - L_t)$ can be also chosen.) The index L_t gives the share of the states of the environment for which the technology has been tested or adapted during the past t times, or the probability that at the next time $(t + 1)$ the environment will take one of its previous values. The index L_t will be called *the maturity level* of the technology or, following the conventional approach of learning models [4], *the learning level* (accordingly, the sequence of its values will be called *the learning curve*). Interestingly, in similar meanings the term "learning curve" is widespread in modern science, starting from the Ebbinghaus "forgetting curves" [5], the psychology of the 20th century (e.g., see the classical papers [6–8] and the monographs [9–11]) and the models suggested by Wright [12] and his followers [13, 14] (the reduced time cost effect of unit production for larger production outputs) and ending with the learning models of artificial neural networks.

In this setup, the technology design and adoption process can be considered from another viewpoint—as the sequential observation of different series of the well-known states of the environment that are interrupted by the newly occurring ones. The length of such a series (from a newly observed state to the next one) has the Bernuolli (binomial) distribution parameterized by the learning level. This parameter is constant during each series and has jumping at the end point of the series when a new state is observed. Then at each time t the expected value of the current series length (till a nearest new state of the environment is observed, exclusive this moment) can be calculated as $L_t (1 - L_t)^{-1}$. The series length corresponds to the number of repetitions required for increasing the maturity (learning) level. In turn, this characterizes, e.g., the time cost of a next learning level increment (in fact, the cost of acquiring new knowledge). Therefore, in some cases *the expected series length* will be employed together with the learning level to describe the learning process. It will be denoted by $N_t = L_t (1 - L_t)^{-1}$, where $L_t < 1$.

A natural formulation of the management problem is to reach a required learning level (a required maturity level of the technology) as fast as possible or using as few resources as possible. Thus, the optimized and/or restricting parameters can be the consumption of resources or time.

In Chap. 1, two technology components management problems have been stated depending on the actor's knowledge about the environment.

The first problem (further referred to as *basic*) rests on the assumption that the list and probabilities of all possible states of the environment are constant and known to the actor (in other words, K and all $\{p_k\}$ are given, independent and time-invariant). This problem surely arises and is basic in the following sense: the actor performs an initial synthesis of the CA technology and testing process (block 3 in Fig. 1.9) using a definite list of all states of the environment. The basic problem is to derive a relation between the technology's maturity level and time as well as to optimize this relation in terms of the available resources.

The second problem is characterized by *the unknown properties of the environment*, i.e., the list and probabilities of all states of the environment (the set K) or at least some of the probabilities $\{p_k\}$ are unknown to the actor or may vary. This problem arises during a continuous repeated use of the technology when (due to the natural variability of the environment) the previously designed technology becomes inadequate in new conditions, which is identified as the result of the productive activities (block 4 in Fig. 2.1). The second problem with the unknown properties of the environment is solved using the laws established for the first problem. Thus, the main attention below will be focused on the first (basic) problem.

Complex technology components are formed by integrating different components within the logical and cause-effect structures of CA [1]. At the general-system level, the logical models of any structural (internally organized) elements of CA are equivalent and have the fan structure; see Chap. 1 of the book for details. At the same time, the cause-effect model is described by *the so-called binary network*, a graph of definite type. Due to the properties of a binary network, it is necessary and sufficient to consider the following integration processes for technology components: (a) sequential integration; (b) parallel conjunctive integration; (c) parallel disjunctive integration. Also a complex integration process that deserves analysis is (d) integration with "the learning to learn" in which a technology component is created simultaneously with its technology. This complex case arises in "pioneering" innovations. For each integration process, the learning level of a complex technology component has to be optimized in terms of time and resources.

Thus, the design and adoption processes of CA technologies generate two management problems for the CA technology components (the basic problem and the problem with unknown properties of the environment) and also four integration problems for the CA technology components (see cases (a)–(d) above), all being embedded in the integrated model (Fig. 1.9).

2.2 Analysis of Design and Adoption of Technology Component

Now, study the properties of *the design and adoption process of a technology component* and also the properties of the learning level L_t in the case where K and all values $\{p_k\}$ are known to the actor of CA.

This process (the row vector x_t) represents a Markov chain with a finite number of states whose numbers y_t are formed from the elements of x_t by the rule $y_t = \sum_{k=1}^{K} x_{kt} 2^{k-1}$. Then the process y_t is also a Markov chain taking any integer values from 0 to $I = \sum_{k=1}^{K} 2^{k-1} = 2^K - 1$ inclusive.

Construct the transition probability matrix $\Pi = \{\pi_{ij}; i = 0, 1, \ldots, I; j = 0, 1, \ldots, I\}$ of the process y_t. At the initial time

$t = 0$, the process y_t is in the 0th state $y_0 = 0$ ($x_{k0} = 0$ for all k) with probability 1. From the state "0," the process may pass only to the states with numbers 2^{k-1} with the probabilities p_k as follows: to the state "1," with the probability p_1; to the state "2," with the probability p_2; to the state "4," with the probability p_3 and so on. The process may not stay in the state "0."

From the state "1," the process may not return to the state "0;" may stay in this state with the probability p_1; and may pass to the state $2^{k-1} + 1$ with the probability p_k for each $1 < k \le K$.

Calculate the elements of the ith row of the matrix Π, where $0 < i \le I$. Consider the binary representation of the number i under the assumption that the first digit is least significant. Denote by $b(i, k)$ the value of the kth digit in this representation. Consequently, $i = \sum_{k=1}^{K} b(i, k)2^{k-1}$.

The probability that the process will stay in the ith state at the next time is $\pi_{ii} = \sum_{k=1}^{K} b(i, k)p_k$. The transition to any other state with a number $j < i$ is impossible ($\pi_{ij} = 0$). For each k such that $x_k = 0$, the probability of transition to the state $i + 2^{k-1}$ is p_k. Finally, the transitions to the other states with the numbers exceeding i are also impossible.

Thus, the elements of the transition probability matrix are given by

$$\pi_{ij} = \begin{cases} 0 & \text{if } j \ne i + 2^n, n = 1, 2, \ldots, K, \\ \sum_{k=1}^{K} b(i, k)p_k & \text{if } j = i, \\ p_n & \text{if } j = i + 2^n, n = 1, 2, \ldots, K \text{ and } j \le I, \end{cases}$$

where $i = 0, 1, \ldots, I$ and $j = 0, 1, \ldots, I$.

Then the transition probability matrix Π of the Markov chain y_t is upper triangular and the state with the maximum number $I = 2^K - 1$ is absorbing: $\pi_{II} = 1$ and $\pi_{Ij} = 0$ for $j \ne I$.

Write the distribution of the state probabilities $q_{it} = \Pr(y_t = i)$ of this chain in the vector notation $q_t = (q_{0t}, q_{1t}, \ldots, q_{it}, \ldots, q_{It})$. At the initial time $t = 0$, the distribution is $q_0 = (1, 0, \ldots, 0)$. Hence, $q_t = q_{t-1}\Pi$ for any $t > 0$ and $q_t = e_0\Pi^t = (1, 0, 0, \ldots, 0)\Pi^t$. Hereinafter, denote by e_i a row vector of appropriate dimension in which all elements are 0 except for the ith one (2.1).

Proposition 1 *As $t \to \infty$, the probability q_{it} of each state with number $0 < i < I$ is majorized by the function $t^{K-1}v^t$, i.e., $q_{it} < \alpha t^{K-1}v^t$, where α and $v < 1$ are some constants.*

The proof of Proposition 1 Divide all states of the Markov chain into groups by the number of unities in their binary representations, i.e., $\sum_{k=1}^{K} b(i, k)$. These groups have two properties, one following from the other.

(1) Any state from the lth group can be reached from the 0th state in l times.
(2) Any state from the lth group can be achieved only from the states of the $(l - 1)$th group.

There exist $K + 1$ groups totally, since l varies from 0 to K. Consider the states from the 1st group; for each of them, $q_{it} = p_k^t$, where k is the number of the corresponding state of the environment.

Then the states of the 1st group satisfy the inequality $q_{it} < \alpha t^0 v_1^t$, where $v_1 = \max_k \{p_k\}$, $\alpha = 1$.

Let $q_{it} < \alpha t^{l-1} v_l^t$ for all states of the lth group. The distribution of the state probabilities of the Markov chain evolves in accordance with the law $q_{it} = \sum_{j=0}^{i-1} \pi_{ji} q_{jt-1} + \pi_{ii} q_{it-1}$. For any state from the $(l + 1)$th group, $q_{it} \equiv 0$ if $t < l$, and $q_{it} = \sum_{\theta=0}^{t-l} \pi_{ii}^\theta \sum_{j=0}^{i-1} \pi_{ji} q_{jt-1-\theta} = \sum_{j=0}^{i-1} \pi_{ji} \sum_{\theta=0}^{t-l} \pi_{ii}^\theta q_{jt-1-\theta}$ if $t \geq l$. This gives the following bound of q_{it}:

$$q_{it} = \sum_{j=0}^{i-1} \pi_{ji} \sum_{\theta=0}^{t-l} \pi_{ii}^\theta q_{jt-1-\theta} < \sum_{j=0}^{i-1} \pi_{ji} \sum_{\theta=0}^{t-l} \pi_{ii}^\theta \alpha (t - 1 - \theta)^{l-1} v_l^{t-1-\theta}$$

$$< \alpha t^{l-1} \sum_{j=0}^{i-1} \pi_{ji} \sum_{\theta=0}^{t-l} \pi_{ii}^\theta v_l^{t-1-\theta} < \alpha t^{l-1} (t - l)(\max\{\pi_{ii}; v_l\})^{t-1} < \alpha v_{l+1}^{-1} t^l v_{l+1}^t,$$

where $v_{l+1} = \max\{v_l; \max_i\{\pi_{ii}\}\}$ and the maximum is calculated over all states i belonging to the $(l + 1)$th group.

Therefore, the inequality $q_{it} < (\alpha v_{l+1}^{-1}) t^l v_{l+1}^t$ holds for all states of the $(l + 1)$th group.

And the desired result follows by mathematical induction for all ith states expect for the Ith one. The proof of Proposition 1 is complete.$\Box$[1]

Proposition 1 has the following practical interpretation: the probabilities of all states of the Markov chain except for the Ith one are decreasing not faster than $t^{K-1} v^t$; conversely, the probability of the Ith state is converging to 1 not slower than $\alpha t^{K-1} v^t$, i.e., $q_{It} > 1 - \alpha t^{K-1} v^t$, where $v = 1 - \min_k \{p_k\}$.

Moreover, Proposition 1 directly implies that there exists the unique stationary distribution $s = (0, 0, \ldots, 0, 1) = e_I$ of the state probabilities of the Markov chain y_t, which is the unique solution of the matrix equation $s = s\Pi$ $(s_i = \sum_{i=0}^{I} \pi_{ji} s_j)$.

Hence, the design and adoption model of technology components described in this section has a unique stable equilibrium—the state in which all possible states of the environment are tested. In other words, any learning level (arbitrary close to 1) can be reached on a sufficiently large horizon.

Now, study *the learning curve*—the behavior of the expected value of the process $L_t = \sum_{k=1}^{K} x_{kt} p_k$. Denote by $E[\cdot]$ the expectation operator.

Above all, in accordance with Proposition 1 the learning level converges in probability to 1:

$$\forall \varepsilon > 0, \lim_{t \to \infty} \{\Pr(|L_t - 1| > \varepsilon)\} = 0.$$

---

[1] Hereinafter, the symbol "$\Box$" indicates the end of a proof or example.

First, by the definition of the process L_t its increments are always nonnegative: $\Delta L_t = L_t - L_{t-1} \geq 0$. In addition, the values of L_t and also the increments ΔL_t are nonnegative and do not exceed 1. Second, the process L_t is also a Markov chain, i.e., L_t and ΔL_t independent random variables for any t. Then[2]

$$E[L_t] = \sum_{k=1}^{K} p_k E[x_{kt}] = \sum_{k=1}^{K} p_k \left(1 - (1 - p_k)^t\right) = 1 - \sum_{k=1}^{K} p_k (1 - p_k)^t. \qquad (2.1)$$

On the other hand, expression (2.1) of $E[L_t]$ can be obtained using the distribution of the state probabilities q_t: $E[L_t] = \sum_{i=1}^{I} \left(\sum_{k=1}^{K} b(i, k) p_k\right) q_{it} = e_0 \Pi^t \beta$, where β denotes a column vector composed of the elements $\sum_{k=1}^{K} b(i, k) p_k$. In turn, they are the diagonal elements of the matrix Π.

From (2.1) the series length can be calculated as

$$N_t = E[L_t](1 - E[L_t])^{-1} = \left(1 - \sum_{k=1}^{K} p_k (1 - p_k)^t\right) \left(\sum_{k=1}^{K} p_k (1 - p_k)^t\right)^{-1}.$$

Since $\Delta L_t = L_t - L_{t-1} \geq 0$, the first differences of the sequence $E[L_t]$ are strictly positive for all t. The formulas of the mth differences, $m \geq 2$, are derived using the $(m - 1)$th differences in the following way:

$$\Delta E[L_t] = 1 - \sum_{k=1}^{K} p_k (1 - p_k)^t - \left(1 - \sum_{k=1}^{K} p_k (1 - p_k)^{t-1}\right)$$

$$= \sum_{k=1}^{K} p_k^2 (1 - p_k)^{t-1} > 0,$$

$$\Delta^2 E[L_t] = \Delta E[L_t] - \Delta E[L_{t-1}] = \sum_{k=1}^{K} p_k^2 (1 - p_k)^{t-1} - \sum_{k=1}^{K} p_k^2 (1 - p_k)^{t-2}$$

$$= - \sum_{k=1}^{K} p_k^3 (1 - p_k)^{t-2} < 0.$$

In the general case,

$$\Delta^m E[L_t] = (-1)^{m+1} \sum_{k=1}^{K} p_k^{m+1} (1 - p_k)^{t-m}.$$

---

[2]Recall that the learning curve L_t describes the probability that the environment will take a new value at time $(t + 1)$. This probability is estimated using the observations during t times inclusive.

Note that, for any time t, the differences of the learning curve form an alternating sequence whose values are decreasing by absolute value ($|\Delta^m E[L_t]| > |\Delta^{m+1} E[L_t]|$). In addition, the first differences satisfy the inequality $1 - E[L_t] > \Delta E[L_t]$.

Thus, the following important result has been established.

Proposition 2 *The learning curve $E[L_t]$ has several properties as follows.*

- *At the initial time $t = 0$, its value is $E[L_0] = 0$.*
- *It is monotonically increasing: $\Delta E[L_t] > 0$.*
- *Its first differences are bounded by the inequality $1 - E[L_t] > \Delta E[L_t]$.*
- *Its growth rate is monotonically decreasing: $\Delta E^2[L_t] < 0$ and $\Delta E^3[L_t] > 0$.*
- *It has asymptotic convergence to 1.*

2.3 Approximation of Learning Curve

Consider some approximations of the learning curve $E[L_t]$ (see Formula 2.1) depending on the probability distribution $P = \left\{ p_k; k = \overline{1, K}; \sum_{k=1}^{K} p_k = 1 \right\}$ of all possible states of the environment.

(A) **Uniform distribution** P. For the sake of simplicity, denote $\delta = 1 / K$. Then

$$E[L_t] = 1 - \sum_{k=1}^{K} p_k (1 - p_k)^t = 1 - \sum_{k=1}^{K} \delta(1 - \delta)^t = 1 - (1 - \delta)^t$$
$$= 1 - \exp(-\gamma t), \tag{2.2}$$

where $\gamma = \ln(1 + 1/(K - 1))$ is the rate of variation of the learning level—*the rate of learning* [1].

The exponential learning curve (2.2) (and its difference analog defined by $E[L_t] = E[L_{t-1}] + \gamma \left(1 - E[L_{t-1}]\right) = \gamma + (1-\gamma)E[L_{t-1}]$) is classical for the theory of learning; see the survey [4] and also the pioneering book [11]. At the same time, for the model under consideration this curve is a special case that corresponds to the uniform distribution of all possible states of the environment. Moreover, as will be demonstrated in Chap. 3, the uniform distribution of all possible states of the environment actually maximizes the expected learning level.

In the uniform distribution case, the expected series length has an exponential growth, $N_t = \exp(\gamma t) - 1$. This is intuitively clear: with further increasing the learning level (and hence the share of the "known" states of the environment), the acquisition of new knowledge requires more efforts "to find" the new states.

The difference equation of N_t has the simple form $N_{t+1} = \exp(\gamma)N_t + (\exp(\gamma) - 1)$; hence, the expected series length is growing multiplicatively.

For $K \gg 1$, the rate of learning becomes

$$\gamma = \ln(1 + 1/(K - 1)) \approx 1/(K - 1) \approx 1/K,$$

and

$$E[L_t] \approx 1 - \exp(-t/K). \tag{2.3}$$

(B) **Distribution (n, δ)** (n highly probable states and $(K - n)$ lowly probable states with $\delta \ll 1/K$). This distribution is given by

$$P = \left\{ p_k = (1 - \delta(K - n))/n, k = \overline{1, n}; p_k = \delta, k = \overline{n + 1, K} \right\}. \tag{2.4}$$

It makes sense to consider the case in which the probabilities $(1 - \delta(K - n))/n$ of the states from the first group are considerably greater than the probabilities δ of the states from the second group. Really, the case in which these probabilities differ insignificantly can be well approximated by the uniform distribution (see above). In other words, let $(1 - \delta(K - n))/n \gg \delta$, which implies $K\delta \ll 1$. Find the learning curve for this distribution:

$$E[L_t] = 1 - \sum_{k=1}^{K} p_k (1 - p_k)^t$$

$$= 1 - \sum_{k=1}^{n} \frac{1 - \delta(K - n)}{n} \left(1 - \frac{1 - \delta(K - n)}{n} \right)^t - \sum_{k=n+1}^{K} \delta(1 - \delta)^t,$$

i.e.,

$$E[L_t] = 1 - (1 - \delta(K - n)) \left(1 - \frac{1}{n} + \frac{(K - n)}{n} \delta \right)^t - (K - n)\delta(1 - \delta)^t.$$

Since $\delta \ll 1/K$ and $n < N$, then $(1 - 1/n) < (1 - \delta)$. Hence, for large t, distribution (2.4) tends to the uniform one:

$$E[L_t] \approx 1 - (K - n)\delta(1 - \delta)^t.$$

For small t, the approximation is

$$E[L_t] \approx 1 - (1 - \delta(K - n)) \left(1 - \frac{1}{n} + \frac{(K - n)}{n} \delta \right)^t.$$

(C) **"Disturbed uniform" distribution**. Let the uniform distribution be disturbed on a "large" set of domains of all possible states of the environment in the following way:

$$P = \left\{ p_k; k = 1, 2, \ldots, K; \sum_{k=1}^{K} p_k = 1; p_k \ll 1; K \gg 1 \right\}. \qquad (2.5)$$

For small t, the learning curve is approximated by

$$E[L_t] = 1 - \sum_{k=1}^{K} p_k (1 - p_k)^t \approx 1 - \sum_{k=1}^{K} p_k (1 - tp_k) = 1 - \sum_{k=1}^{K} p_k + t \sum_{k=1}^{K} p_k^2$$

$$= t \sum_{k=1}^{K} p_k^2, \qquad (2.6)$$

i.e., has a linear growth in t with the rate $\sum_{k=1}^{K} p_k^2$.

For large t, there are two possible behavior patterns of the learning curve as follows. If all domains of all possible states of the environment are nearly equivalent and their probabilities differ insignificantly ($p_k \approx 1/K, k = \overline{1, K}$), then the uniform distribution estimate (2.3) also holds. If a certain number n of the domains have a substantial distinction from the others, then an adequate description is the approximation $E[L_t] \approx 1 - \left(1 - \frac{n}{K}\right)\left(1 - \frac{1}{K}\right)^t$ (distribution B).

Interestingly, the analytical expressions (2.2), (2.3) and (2.6) as well as the properties of the learning curve $E[L_t]$ established within the current model well match many conventional models of learning (in particular, the ones discussed in Sects. 4.2 and 6.7 of the book [4]). However, the well-known models postulate the form of the learning curve or its equations, whereas the model suggested in this section describes the process of learning (design) and the equations and properties of the learning curve are derived during model analysis.

2.4 Expected Learning Time

In this section, the expected time of reaching *a required learning level* $L_{\text{req}} \in (0, 1)$ (the technology's maturity level) will be calculated. This is the expected time t at which

$$L_t = \sum_{k=1}^{K} x_{kt} p_k \geq L_{\text{req}}.$$

For this purpose, study the behavior of the Markov chain y_t, in particular, the evolution of the probability distribution $q_{it} = Pr(y_t = i)$ of its states. As it has been established earlier, the initial distribution for $t = 0$ is $q_0 = (1, 0, \ldots, 0)$ and also $q_t = q_{t-1} \Pi$ for any $t > 0$, $q_t = e_0 \Pi^t = (1, 0, 0, \ldots, 0) \Pi^t$.

The matrix Π is upper triangular, which gives several properties of the matrix Π^t as follows.

- The determinant of the matrix Π (denoted by $\Delta\Pi$) is the product of all its diagonal elements: $\Delta\Pi = \prod_{i=1}^{I} \pi_{ii}$.
- The matrix Π^t is also an upper triangular. (This fact is immediate from the multiplication rules of matrices.)
- The diagonal elements of the matrix Π^t are the powers of the diagonal elements of the matrix Π: $\pi_{ii}^t = (\pi_{ii})^t$.
- The determinant of the matrix Π^t is the product of the determinant Π: $\Delta\Pi^t = (\Delta\Pi)^t = \left(\prod_{i=1}^{I} \pi_{ii}\right)^t$.

Construct a mask—a column vector r of the same dimension as the row vector q_t—by the rule

$$r_i = \begin{cases} 1 \text{ if } \sum_{k=1}^{K} b(i,k)p_k < L_{\text{req}}, \\ 0 \text{ if } \sum_{k=1}^{K} b(i,k)p_k \geq L_{\text{req}}, \end{cases}$$

where $i = \sum_{k=1}^{K} 2^k b(i,k) = 0, 1, \ldots, I$.

This mask vector "extracts" the states of the process y_t for which the learning levels are below the required one. Then, for each time, the probability that the learning level has reached or exceeded the required level is $\Pr(L_t \geq L_{\text{req}}) = 1 - q_t r$ (equivalently, $\Pr(L_t < L_{\text{req}}) = q_t r$). The probability that the time t_{reach} of reaching the required learning level exceeds the current time is $\Pr(t_{\text{reach}} > t) = q_t r$; the probability that the required learning level has been reached by the current time is $\Pr(t_{\text{reach}} \leq t) = 1 - q_t r$. Obviously, $r_I = 0$. In accordance with Proposition 1, the probabilities $\Pr(L_t < L_{\text{req}}) = \Pr(t_{\text{req}} > t)$ can be majorized by the function $t^{K-1} v^t$ as $t \to \infty$. Consequently,

$$\Pr(t_{\text{reach}} > t) = \sum_{i=0}^{I} r_i q_t < \sum_{i=0}^{I} r_i \alpha t^{K-1} v^t = \left(\alpha \sum_{i=0}^{I} r_i\right) t^{K-1} v^t.$$

Using the relation $q_t = e_0 \Pi^t$, write $\Pr(t_{\text{reach}} > t) = q_t r = e_0 \Pi^t r$. Since $\Pr(t_{\text{reach}} > t - 1) = \Pr(t_{\text{reach}} = t) + \Pr(t_{\text{reach}} > t)$, it follows that $\Pr(t_{\text{reach}} = t) = \Pr(t_{\text{reach}} > t - 1) - \Pr(t_{\text{reach}} > t)$.

Then the expected time can be calculated as

$$\bar{t}_{\text{reach}} = \sum_{t=0}^{\infty} t \Pr(t_{\text{reach}} = t) = \sum_{t=0}^{\infty} t(\Pr(t_{\text{reach}} > t - 1) - \Pr(t_{\text{reach}} > t)).$$

Because all probabilities $\Pr(t_{\text{reach}} > t)$ are majorized by $t^{K-1}v^t$ as $t \to \infty$, the series $t\Pr(t_{\text{reach}} > t)$ is converging and has a finite sum $\sum_{t=0}^{\infty} t \Pr(t_{\text{reach}} > t)$. As a result,

$$\bar{t}_{\text{reach}} = \sum_{t=0}^{\infty} t(\Pr(t_{\text{reach}} > t - 1) - \Pr(t_{\text{reach}} > t))$$

$$= \sum_{t=0}^{\infty} (t+1)\Pr(t_{\text{reach}} > t) - \sum_{t=0}^{\infty} t\Pr(t_{\text{reach}} > t) = \sum_{t=0}^{\infty} \Pr(t_{\text{reach}} > t).$$

Thus,

$$\bar{t}_{\text{reach}} = \sum_{t=0}^{\infty} \Pr(t_{\text{reach}} > t) = \sum_{t=0}^{\infty} e_0 \Pi^t r = e_0 \left(\sum_{t=0}^{\infty} \Pi^t \right) r = e_0(E - \Pi)^{-1}r, \quad (2.7)$$

where E is an identity matrix of the same dimension as the matrix Π. The existence of the inverse $(E - \Pi)^{-1}$ follows from the upper triangular property of the matrix $(E - \Pi)$. All its diagonal elements can be found from the above expression of π_{ij}, and their product is positive.

This theoretical development naturally leads to the following result.

Proposition 3 *For any Markov chain, the expected time of first reaching a state from the given set is $\bar{t}_{\text{reach}} = \sum_{t=0}^{\infty} \Pr(t_{\text{reach}} > t)$. If this series is converging, then $\bar{t}_{\text{reach}} = e_0(E - \Pi)^{-1}r$.*

Generally speaking, Formula (2.7) can be used for calculating $\bar{t}_{\text{reach}}$ depending on the probability distribution $P = \{p_k; k = 1, 2, \ldots, K\}$ of all possible states of the environment (as the values p_k are taken into account through the matrix Π) and also on the required learning level (as the value L_{req} is taken into account through the vector m). Unfortunately, Formula (2.7) is not constructive because neither the probabilities p_k nor the level L_{req} enter it in explicit form. Furthermore, the considerable dimensions of the matrix Π $(2^K \times 2^K)$ make the use of (2.7) difficult in practice.

In a series of special cases, simpler and more constructive expressions can be obtained. Consider one of them—the uniform distribution $P = \{p_k = 1/K = \delta; k = \overline{1, K}\}$ of all possible states of the environment. In this case, instead of the Markov chain y_t with the values $i = 0, 1, \ldots, I$, consider a chain $\tilde{y}_t$ whose values correspond to the number of the states of the environment for which the technology has been tested. In other words, the chain $\tilde{y}_t$ takes the values from 0 to K inclusive. Then $L_t = \delta \tilde{y}_t$, and the transition probabilities have the form

$$\pi_{ij} = \begin{cases} 0 & \text{if } j < i \text{ or } j > i+1, \\ 1 - \delta i & \text{if } j = i+1, \\ \delta i & \text{if } j = i, \end{cases}$$

where $i = 0, 1, \ldots, K$.

Then the matrix $(E - \Pi)$ is an upper triangular, bad matrix with the elements

$$\varepsilon_{ij} = \begin{cases} 0 & \text{if } j < i \text{ or } j > i+1, \\ 1 - \delta i & \text{if } j = i, \\ \delta i - 1 & \text{if } j = i+1, \end{cases}$$

where $i = 0, 1, \ldots, K$.

The inverse $(E - \Pi)^{-1}$ is also an upper triangular matrix with the elements

$$\varepsilon_{ij}^{-} = \begin{cases} 0 & \text{if } j < i, \\ (1 - \delta i)^{-1} & \text{if } i \le j < K, \\ 1 & \text{if } j = K, \end{cases}$$

where $i = 0, 1, \ldots, K$.

The operation $(1, 0, 0, \ldots, 0)(E - \Pi)^{-1} r$ removes the first row from the matrix $(E - \Pi)^{-1}$ and sums up those elements of this row for which $\sum_{k=1}^{K} x_k p_k < L_{\text{req}}$. In the uniform distribution case, the sum consists of the first $L_{\text{req}}/\delta = K L_{\text{req}}$ elements, and the expected time of reaching the required learning level L_{req} is

$$\bar{t}_{\text{reach}} = \sum_{i=0}^{K L_{\text{req}}} (1 - \delta i)^{-1} = \sum_{i=0}^{K L_{\text{req}}} \frac{K}{K - i}.$$

For $K \gg 1$, the value $\bar{t}_{\text{reach}}$ has the compact approximation

$$\bar{t}_{\text{reach}} = \sum_{i=0}^{K L_{\text{req}}} \frac{K}{K - i} = K \sum_{i=0}^{K L_{\text{req}}} \frac{1}{1 - K^{-1} i} K^{-1} \approx K \int_{0}^{K L_{\text{req}}} \frac{1}{1 - K^{-1} x} K^{-1} dx =$$

$$= -K \ln\left(1 - K^{-1} x\right)\Big|_{0}^{K L_{\text{req}}} = -K \ln\left(1 - L_{\text{req}}\right),$$

and

$$\bar{t}_{\text{reach}} \approx -K \ln\left(1 - L_{\text{req}}\right). \tag{2.8}$$

Note that expression (2.8) coincides with the approximate solution $\hat{t}$ of the equation $E[L_t] = 1 - (1 - \delta)^{\hat{t}} = L_{\text{req}}$ (see (2.2)), for which $\hat{t} = \frac{\ln(1 - L_{\text{req}})}{\ln(1 - \delta)} \approx -\frac{\ln(1 - L_{\text{req}})}{\delta} = -K\ln\left(1 - L_{\text{req}}\right)$.

Next, calculate the expected time of reaching the "absolute" learning level $L_t = 1$.

Assume several states of the environment have been tested by some time; let $I \leq K$ states with the probabilities $\{p_i; i = \overline{1,I}\}$ be still unknown. Clearly, $\sum_{i=1}^{I} p_i \leq 1$.

Denote by $T(\{p_i; i = \overline{1,I}\})$ the expected time of testing the residual I states, i.e., the expected time of reaching the learning level $L_t = 1$. Using mathematical induction, prove the formula

$$T(\{p_i; i = \overline{1,I}\}) = \sum_{i=1}^{I} p_i^{-1} - \sum_{i;k} (p_i + p_k)^{-1} + \sum_{i;k;l} (p_i + p_k + p_l)^{-1} - \ldots \quad (2.9)$$

Write (2.9) in another (equivalent) form:

$$T(\{p_i; i = \overline{1,I}\}) = \sum_{k=1}^{I} (-1)^{k+1} \sum_{i_1;i_2;\ldots;i_k} \left(\sum_{j=1}^{k} p_{i_j}\right)^{-1}. \quad (2.10)$$

In other words, it is necessary to demonstrate that the expected time represents the sum of the alternating partial sums of all ks from $\{p_i; i = \overline{1,I}\}, n = \overline{1,I}$.

Introduce the compact notation $\Theta(k; I)$ for the sums of all ks from $\{p_i; i = \overline{1,I}\}$, $k = \overline{1,I}$, that appear in (2.10):

$$\Theta(k; I) = \sum_{i_1;i_2;\ldots;i_k} \left(\sum_{j=1}^{k} p_{i_j}\right)^{-1} = \sum_{i_1;i_2;\ldots;i_k} \frac{1}{p_{i_1} + p_{i_2} + \ldots + p_{i_k}}. \quad (2.11)$$

In view of (11), expression (2.10) takes the form

$$T(\{p_i; i = \overline{1,I}\}) = \sum_{k=1}^{I} (-1)^{k+1} \sum_{i_1;i_2;\ldots;i_k} \left(\sum_{j=1}^{k} p_{i_j}\right)^{-1} = \sum_{j=1}^{I} (-1)^{j+1} \Theta(j; I).$$

$$(2.12)$$

Assume a single state of the environment remains untested; then $T(\{p_i; i = 1\}) = 1/p_1$.

Let relations (2.9), (2.10) and (2.12) be valid for the $(I - 1)$ states, i.e., $T(\{p_i; i = \overline{1, I-1}\}) = \sum_{k=1}^{I-1} (-1)^{k+1} \Theta(k; I-1)$.

Show that, in this case, relations (2.9), (2.10) and (2.12) will be valid for the I states. The event that the $i = \overline{1,I}$ states have been realized sequentially is the union of the I events in each of which one of the I states has been realized first and the other $(I - 1)$ states sequentially after it. Then

$$T(\{p_i; i = \overline{1,I}\}) = \left(\sum_{i=1}^{I} p_i\right)^{-1} + \sum_{j=1}^{I} p_j \left(\sum_{i=1}^{I} p_i\right)^{-1} T(\{p_i; i = \overline{1,I}; i \neq j\}),$$

(2.13)

where the first term is the expected time to the first of the realized states $i = 1, 2, \ldots$ I. Each of the jth terms in the summand gives the probability that the jth state has been tested first; then $T(\{p_i; i = \overline{1,I}; i \neq j\})$ gives the expected testing time of the residual $(I - 1)$ states.

Substituting (2.12) into the right-hand side of (2.13) yields

$$T(\{p_i; i = \overline{1,I}\}) = \left(\sum_{i=1}^{I} p_i\right)^{-1} + \left(\sum_{i=1}^{I} p_i\right)^{-1} \sum_{j=1}^{I} \left(p_j \sum_{k=1}^{I-1} (-1)^{k+1} \Theta(k; I - 1)\right). \quad (2.14)$$

Using (2.11) transform the second sum in the following way:

$$\sum_{j=1}^{I} \left(p_j \sum_{k=1}^{I-1} (-1)^{k+1} \Theta(k; I - 1)\right) = \sum_{j=1}^{I} \sum_{k=1}^{I-1} (-1)^{k+1} \sum_{i_1; i_2; \ldots; i_k} \frac{p_j}{p_{i_1} + p_{i_2} + \ldots + p_{i_k}}.$$

Each of the probabilities p_j in the numerators does not appear in the sum $p_{i_1} + p_{i_2} + \cdots + p_{i_k}$ in the denominators. Hence, the order of summation can be modified so that

$$\sum_{j=1}^{I} \left(p_j \sum_{k=1}^{I-1} (-1)^{k+1} \Theta(k; I - 1)\right)$$

$$= \sum_{j=1}^{I} \sum_{k=1}^{I-1} (-1)^{k+1} \sum_{i_1; i_2; \ldots; i_k} \frac{p_j}{p_{i_1} + p_{i_2} + cdots + p_{i_k}}$$

$$= \sum_{k=1}^{I-1} (-1)^{k+1} \sum_{i_1; i_2; \ldots; i_k} \left(p_{i_1} + p_{i_2} + \cdots + p_{i_k}\right)^{-1} \sum_{j=1; j \neq i_1; i_2; \ldots; i_k}^{I} p_j.$$

Here the summation procedure runs over all js from 1 to I not coinciding with any of $i_1, i_2, \ldots, i_k$. Consequently,

$$\sum_{j=1; j \neq i_1; i_2; \ldots; i_k}^{I} p_j = \left(\sum_{i=1}^{I} p_i\right) - \left(p_{i_1} + p_{i_2} + \cdots + p_{i_k}\right).$$

Using this relation in the second sum gives

$$\sum_{j=1}^{I}\left(p_j \sum_{k=1}^{I-1}(-1)^{k+1}\Theta(k; I-1)\right)$$

$$= \sum_{k=1}^{I-1}(-1)^{k+1}\sum_{i_1;i_2;\dots;i_k}\frac{\left(\sum_{i=1}^{I}p_i\right)-(p_{i_1}+p_{i_2}+\cdots+p_{i_k})}{p_{i_1}+p_{i_2}+\cdots+p_{i_k}}$$

$$= \left(\sum_{i=1}^{I}p_i\right)\sum_{k=1}^{I-1}(-1)^{k+1}\sum_{i_1;i_2;\dots;i_k}(p_{i_1}+p_{i_2}+\cdots+p_{i_k})^{-1}-\sum_{k=1}^{I-1}(-1)^{k+1}\sum_{i_1;i_2;\dots;i_k}1$$

$$= \left(\sum_{i=1}^{I}p_i\right)\sum_{k=1}^{I-1}(-1)^{k+1}\Theta(k; I)+\sum_{k=1}^{I-1}(-1)^{k}C_I^k,$$

where C_I^k is the number of k-combinations from the set of I elements.
Substituting this formula into (2.14) yields

$$T\left(\{p_i; i = \overline{1,I}\}\right) = \left(\sum_{i=1}^{I}p_i\right)^{-1} + \left(\sum_{i=1}^{I}p_i\right)^{-1}\left(\left(\sum_{i=1}^{I}p_i\right)\sum_{k=1}^{I-1}(-1)^{k+1}\Theta(k; I)+\right.$$

$$= \sum_{k=1}^{I-1}(-1)^{k+1}\Theta(k; I) + \left(\sum_{i=1}^{I}p_i\right)^{-1}\left(1 + \sum_{k=1}^{I-1}(-1)^{k}C_I^k\right).$$

By definition, $\Theta(I; I) = \left(\sum_{i=1}^{I}p_i\right)^{-1}$, and also $1 + \sum_{n=1}^{I-1}(-1)^n C_I^n + (-1)^I = (1-1)^I = 0$. Hence, $1 + \sum_{n=1}^{I-1}(-1)^n C_I^n = (-1)^{I+1}$.

Finally, the desired expected time of reaching the "absolute" learning level $L_t = 1$ (in form 2.12) is

$$T\left(\{p_i; i = \overline{1,I}\}\right) = \sum_{k=1}^{I}(-1)^{k+1}\Theta(k; I).$$

The outcomes of Sects. 2.2–2.4 can be summarized as follows. The properties of the design and adoption process of a technology component, the learning level and the expected learning time have been studied in detail. The next stage is to analyze the integration models of technology components, which will be done in the next section.

2.5 Integration Models of Technology Components

Consider the *integration* of partial *technology components*, each described by the basic model, as the following processes:

(A) sequential integration;
(B) parallel conjunctive integration;
(C) parallel disjunctive integration;
(D) parallel integration with complete information exchange;
(E) integration with "the learning to learn."

    To examine their properties, consider the management process of several technology components with an appropriate integration of their results. The states of partial processes are Markov chains with the properties established above. Then an integrated process will also evolve as a Markov chain on the state set defined by the direct product of the state sets of the partial processes.

    Following the same approach as before, introduce a $\tilde{K}$-dimensional process $\tilde{x}_t = \left(x_{1t}, x_{2t}, \ldots, x_{kt}, \ldots, x_{\tilde{K}t}\right)$, $\tilde{K} = \sum_{m=1}^{M} K^m$, in which each element x_{kt} takes value 0 or 1, meaning that a corresponding state of the environment is untested or tested, respectively. Also introduce a process $\tilde{y}_t$ that reflects the number of a current state of the process $\tilde{x}_t$. Both processes $\tilde{x}_t$ and $\tilde{y}_t$ are Markov chains. The transition probability matrix of the process $\tilde{y}_t$ is upper triangular, and this process satisfies Proposition 1 (on the asymptotic behavior of the probability distribution of states) and Proposition 3 (on the expected time of reaching a given learning level).

A. Let the integration process be intended to create all partial technology components (the conjunction of all M partial components). Then the learning level $L_t^{1\ldots M}$ (the maturity level of the integrated technology) is the probability that none of the partial processes will have an untested state of the environment during a next test—the product of the maturity levels of all partial technologies. In turn, this probability is the product of the probabilities L_t^m of all partial technology components: $L_t^{1\ldots M} = \prod_{m=1}^{M} L_t^m$. Consequently,

$$E\left[L_t^{1\ldots M}\right] = \prod_{m=1}^{M} E\left[L_t^m\right] = \prod_{m=1}^{M} \left[1 - \sum_{k=1}^{K} p_k^m \left(1 - p_k^m\right)^t\right]. \qquad (2.15)$$

    Formula (2.7) that determines the expected time of reaching a required learning level of a single technology component can be easily extended to the case of M elements as follows:

$$\bar{t}_{A\,\text{reach}} = \sum_{t=0}^{\infty} \Pr(t_{A\,\text{reach}} > t) = \sum_{t=0}^{\infty} \Pr\left(\max_{m}\{t_{m\,\text{reach}}\} > t\right)$$

$$= \sum_{t=0}^{\infty} \left(1 - \prod_{m=1}^{M} \left(1 - e_0 \Pi_m^t r_m\right)\right). \qquad (2.16)$$

If the design processes of all technology components have the same characteristics, then

$$\bar{t}_{A\,\text{reach}} = \sum_{t=0}^{\infty}\left(1 - (1 - e_0\Pi^t r)^M\right) = \sum_{t=0}^{\infty}\left(\sum_{m=1}^{M} C_M^m (-1)^{m-1}(e_0\Pi^t r)^m\right)$$

$$= M\sum_{t=0}^{\infty} e_0\Pi^t r + \sum_{t=0}^{\infty}\left(\sum_{m=2}^{M-1} C_M^m (-1)^{m-1}(e_0\Pi^t r)^m\right) + (-1)^{M-1}\sum_{t=0}^{\infty}(e_0\Pi^t r)^M.$$

Using the sequence of the expected times $\bar{t}_{A\,\text{reach}}$ for different increasing values M, we can calculate the first differences $\Delta\bar{t}_M = \sum_{t=0}^{\infty} e_0\Pi^t r(1 - e_0\Pi^t r)^{M-1}$ and also the second differences $\Delta^2\bar{t}_M = -\sum_{t=0}^{\infty}(e_0\Pi^t r)^2(1 - e_0\Pi^t r)^{M-2}$. Clearly, $\bar{t}_{A\,\text{reach}}$ is growing with M but the rate of growth is an increasing function of M. In addition, the first differences are bounded above and below:

$$\bar{t}_{\text{reach}} - \sum_{t=0}^{\infty}(e_0\Pi^t r)^2 = \sum_{t=0}^{\infty} e_0\Pi^t r(1 - e_0\Pi^t r) \leq \Delta\bar{t}_M < \sum_{t=0}^{\infty} e_0\Pi^t r = \bar{t}_{\text{reach}}.$$

B. Let all partial processes be independently implemented in parallel to each other and also let the integration process be intended to create at least one of the partial components (the disjunction of M partial components). Then the "non-maturity" level of the complex technology is the share of the untested states of the complex environment, $1 - L_t^{1\_M} = \prod_{m=1}^{M}(1 - L_t^m)$ and hence

$$E[L_t^{1\_M}] = 1 - \prod_{m=1}^{M}\left[\sum_{k=1}^{K} p_k^m(1 - p_k^m)^t\right]. \tag{2.17}$$

In this case, the expected time of reaching a required learning level of M elements is calculated as

$$\bar{t}_{B\,\text{reach}} = \sum_{t=0}^{\infty}\Pr(t_{B\,\text{reach}} > t) = \sum_{t=0}^{\infty}\Pr\left(\min_m\{t_{m\,\text{reach}}\} > t\right) = \sum_{t=0}^{\infty}\left(\prod_{m=1}^{M} e_0\Pi_m^t r_m\right). \tag{2.18}$$

If the partial processes have the same characteristics, then

$$\bar{t}_{B\,\text{reach}} = \sum_{t=0}^{\infty}\left(\prod_{m=1}^{M} e_0\Pi_m^t r_m\right) = \sum_{t=0}^{\infty}(e_0\Pi^t r)^M.$$

For the parallel implementation of several partial processes with the same characteristics (cases A and B), the expected time is

$$\bar{t}_{A \text{ reach}} = M\bar{t}_{\text{reach}} + \sum_{t=0}^{\infty} \left(\sum_{m=2}^{M-1} C_M^m (-1)^{m-1} \left(e_0 \Pi^t r\right)^m \right) + (-1)^{M-1} \bar{t}_{B \text{ reach}},$$

where $\bar{t}_{\text{reach}}$, $\bar{t}_{A \text{ reach}}$ and $\bar{t}_{B \text{ reach}}$ denote the expected times of completing a partial process, all M partial processes and at least one of the M partial processes, respectively. In addition, the following chain of bounds holds:

$$M\bar{t}_{\text{reach}} - (M-1)\bar{t}_{B \text{ reach}} \le \bar{t}_{A \text{ reach}} < M\bar{t}_{\text{reach}}.$$

For the parallel implementation of two partial processes with the same characteristics (cases A and B), the expected time formula gives $\bar{t}_{A \text{ reach}} = 2\bar{t}_{\text{reach}} - \bar{t}_{B \text{ reach}}$. Hence, the expected times are related by

$$\bar{t}_{\text{reach}} = \left(\bar{t}_{A \text{ reach}} + \bar{t}_{B \text{ reach}}\right)/2.$$

C. Let two technology components be implemented sequentially so that the second component is initiated directly after the completion of the first component. This case is described by two independent Markov chains: the second chain starts evolution from a known state as soon as the state of the first chain reaches a given domain. Such a complex technology consists of two elements and the second element can be designed only after the completion of the first component. The probability distribution of the design completion time of this complex technology—the time of reaching a given domain for the second chain—is the convolution of the probability distributions of the times for both chains. This law can be used for calculating the integrated learning curve and the expected design time as the sum of the expected design times of the partial technologies.

D. Let a technology component be independently implemented in parallel within several (M) processes with complete information exchange. Then M independent tests are organized during one period (between two successive times) and hence

$$E\left[L_t^{1||M}\right] = 1 - \sum_{k=1}^{K} p_k (1 - p_k)^{Mt}. \tag{2.19}$$

In this case, the expected time of reaching a required learning level of M elements can be calculated as

$$\bar{t}_{\text{reach}} = \sum_{t=0}^{\infty} \Pr(t_{\text{reach}} > t) = \sum_{t=0}^{\infty} e_0 \Pi^{mt} r = e_0 \left(E - \Pi^m\right)^{-1} r. \tag{2.20}$$

As is easily demonstrated, the learning curves (2.15), (2.17) and (2.19) satisfy all statements of Proposition 2.

E. **"Learning to learn."** Let the process of technology adoption be running simultaneously with its design. In other words, assume the intensity of environment testing is varying in accordance with another learning curve.

Consider a process $L_t = \sum_{k=1}^{K} x_{kt} p_k$ that describes the design and adoption process of a new technology and also a process $\tilde{L}_t = \sum_{j=1}^{J} \tilde{x}_{jt} q_j$ that describes the management process of the design technology of the former technology (*"the learning to learn"*). The processes L_t and $\tilde{L}_t$ will be assumed to be statistically independent.

At each time t, the states of the environment are tested with the probability $\tilde{L}_t$ or "skipped" with the probability $(1 - \tilde{L}_t)$. In the latter case, a state of the environment is not tested and the processes x_{kt} has invariable states.[3] Consequently,

$$
E\big[x_{kt+1}|x_{kt}\big] = \begin{cases} x_{kt} & \text{with probability } 1 - \sum_{j=1}^{J} \tilde{x}_{jt} q_j, \\ x_{kt} + (1 - x_{kt}) p_k & \text{with probability } \sum_{j=1}^{J} \tilde{x}_{jt} q_j, \end{cases}
$$

where $E[\cdot|x_{kt}]$ denotes the conditional expectation operator given x_{kt}.

Then $E\big[x_{kt+1}|x_{kt}\big] = x_{kt} + \Gamma_t p_k (1 - x_{kt})$, where $\Gamma_t = E\big[\sum_{j=1}^{J} q_j \tilde{x}_{jt}\big] = 1 - \sum_{j=1}^{J} q_j (1 - q_j)^t$.

Passing from the conditional to unconditional expectations yields the difference equation

$$
E\big[x_{kt+1}\big] = E[x_{kt}] + \Gamma_t p_k (1 - E[x_{kt}]), \tag{2.21}
$$

which can be used for calculating $E[x_{kt}]$ sequentially for all $t \geq 0$.

Introduce the notation $\Psi_t = 1 - E[x_{kt}]$. Then
$E\big[x_{kt+1}\big] = 1 - \Psi_{t+1} = \Psi_t \Gamma_t p_k + 1 - \Psi_t$ and $\Psi_{t+1} = \Psi_t (1 - \Gamma_t p_k)$.
Since $\Psi_0 = 1$, it follows that $\Psi_t = \prod_{\tau=0}^{t-1} (1 - \Gamma_\tau p_k)$ and consequently

$$
E[x_{kt}] = 1 - \prod_{\tau=0}^{t-1} (1 - \Gamma_\tau p_k) = 1 - \prod_{\tau=0}^{t-1} \left(1 - p_k + p_k \sum_{j=1}^{J} q_j (1 - q_j)^\tau \right).
$$

Finally, the learning curve takes the form

---

[3] This model may have an alternative interpretation as follows. Checks are performed at each step while a technology for a new state is designed with some probability determined by a metaprocess. In the logistic model, this probability is equal to the learning level in the process itself; in the hyperbolic model, to the probability of "error" raised to some power with the proportionality factor μ.

$$E[L_t] = \sum_{k=1}^{K} p_k E[x_{kt}] = \sum_{k=1}^{K} p_k \left(1 - \prod_{\tau=0}^{t-1} (1 - \Gamma_\tau p_k) \right)$$

$$= 1 - \sum_{k=1}^{K} p_k \prod_{\tau=0}^{t-1} \left(1 - p_k + p_k \sum_{j=1}^{J} q_j (1 - q_j)^\tau \right). \quad (2.22)$$

Consider expression (2.21) in detail. Calculate the first differences: $\Delta E[x_{kt}] = \Gamma_t p_k \Psi_t$. Note that (a) $\Delta E[x_{kt}]|_{t=0} = 0$ for any p_k because $\Gamma_0 = 0$ and (b) $\Delta E[x_{kt}] > 0$ for any $t > 0$. In other words, $E[x_{kt}]$ is growing for $t > 1$, which seems intuitively clear.

Calculate the second differences:

$$\Delta^2 E[x_{kt}] = \Delta E[x_{kt+1}] - \Delta E[x_{kt}] = \Gamma_{t+1} p_k (\Psi_t - \Psi_t \Gamma_t p_k) - \Gamma_t p_k \Psi_t$$

$$= \Gamma_{t+1} p_k (1 - \Gamma_t p_k) \Psi_t - \Gamma_t p_k \Psi_t = p_k \Psi_t (\Gamma_{t+1}(1 - \Gamma_t p_k) - \Gamma_t) \quad (2.23)$$

Above all, $\Delta^2 E[x_{kt}]|_{t=0} = p_k \Gamma_1 > 0$. However, Γ_t is monotonically increasing from 0 and asymptotically converging to 1 as $t \to \infty$. Then $\Delta^2 E[x_{kt}]|_{t \to \infty} = -p_k^2 \Psi_t < 0$ and $\Delta^2 E[x_{kt}]|_{t \to \infty} \to 0-$.

The learning curve $E[L_t] = \sum_{k=1}^{K} p_k E[x_{kt}]$ is a linear combination of the processes $E[x_{kt}]$ with strictly positive coefficients. Hence, the first and second differences of the learning curve satisfy all statements formulated for $E[x_{kt}]$. More specifically,

- $E[L_t]|_{t=0} = 0$;
- $\Delta E[L_t]|_{t=0} = 0$ and $\Delta E[L_t] > 0$ for all $t > 0$ (the learning curve is increasing in t from 0 and asymptotically converging to 1);
- $\Delta^2 E[L_t]|_{t=0} > 0$, $\Delta^2 E[L_t]|_{t \to \infty} < 0$ and $\Delta^2 E[L_t]|_{t \to \infty} \to 0-$ (the learning curve has an *inflection point*, being strictly convex on the left and strictly concave on the right of it).

Consider <u>an example</u> for "the learning to learn." Let all possible states of the environment be uniformly distributed, $P = \{p_k = 1/K; \ k = \overline{1, K}\}$, and also let $Q = \{q_j = 1/J; \ j = \overline{1, J}\}$. Using the notation $\eta = J^{-1}$, write

$$G_t = E\left[\sum_{j=1}^{J} q_j \tilde{x}_{jt} \right] = 1 - \sum_{j=1}^{J} J^{-1} (1 - J^{-1})^t = 1 - (1 - J^{-1})^t = 1 - (1 - \eta)^t.$$

The corresponding learning curve has the form

$$E[L_t] = 1 - \prod_{\tau=0}^{t-1} (1 - \delta(1 - (1 - \eta)^\tau)). \quad (2.24)$$

The second difference Formula (2.17) can be employed for estimating the inflection point $\hat{t}$ of the learning curve (2.18) in the case of complex technologies ($K \gg 1$, $J \gg 1$ and $K < J$). The resulting estimate is

$$\hat{t} \approx \sqrt{KJ + 0.25K^2} - 0.5K.$$

Write (2.24) in the equivalent form $E[L_t] = 1 - \exp\left[\sum_{\tau=0}^{k-1} \ln(1 - \delta \exp(-\varphi\tau))\right]$, where $\varphi = \ln(1 + \frac{1}{J-1})$. Then

$$E[L_t] = E[L_{t-1}] + \delta(1 - (1-\eta)^{t-1}) \prod_{\tau=0}^{k-2} (1 - \delta(1 - (1-\eta)^{\tau}))$$

$$= E[L_{t-1}] + (1 - E[L_{t-1}])\delta(1 - (1-\eta)^{t-1}) = \beta_t + (1 - \beta_t)E[L_{t-1}],$$

where $\beta_t = \delta(1 - (1-\eta)^{t-1}) = \delta(1 - \exp(-\varphi(t-1)))$. Thus, the learning curve (2.24) is a "generalization" of the learning curve (2.2) in which the coefficients $\{\beta_t\}$ of the difference equation depend on the time variable.

Consider several special cases of learning to learn, namely, processes in which the probability of successful technology design for each of the first encountered states of the environment is not identically 1 and depends not on the state of a similar "external" process (see expression 2.22) but on the learning level reached in the process itself. Moreover, this relation can be both increasing (see the logistic learning curve model below) and decreasing (see the hyperbolic learning curve model below). The corresponding class of learning processes will be conventionally called *auto-learning*.

Logistic learning curve. Consider an important special case of "learning to learn" as follows. For a sufficiently large number of practical situations, the intensity of testing different states of the environment is proportional to the learning level: $\Gamma_t = \mu E[L_t]$. Really, when testing new (e.g., aerospace or transport) equipment, or commissioning new production complexes, at the first stages the product or complex is often tested for a limited set of operating modes (bench and ground tests, idling, etc.). As experience develops, the range of modes is expanded to the complete set of all possible modes and conditions of the environment, and the transition to standard use is performed, which well matches the formal assumption that "the rate of learning" is proportional to the reached level.

This situation corresponds to a special case of the self-learning model in which the design process $\tilde{L}_t$ of the technology coincides with its adoption process L_t.

For the sake of analysis, rewrite the difference Eq. (2.21) in a slightly modified form as follows:

$$\Delta E[x_{kt+1}] = \Gamma_t p_k (1 - E[x_{kt}]). \tag{2.25}$$

If all states of the environment are equally probable ($p_k = \delta = 1/K$), then due to (2.25) the learning level satisfies the difference equation

$$\delta \, \Delta E[x_{kt+1}] = \Gamma_t \delta^2 (1 - E[x_{kt}]) = \Gamma_t \delta^2 - \Gamma_t \delta (\delta E[x_{kt}]).$$

Summation over k yields

$$\Delta E[L_{t+1}] = \Gamma_t K \, \delta^2 - \Gamma_t \delta \, E[L_t] = \Gamma_t \delta (1 - E[L_t]). \tag{2.26}$$

In the case $\Gamma_t = \mu \, E[L_t]$, the difference equation of the learning level takes the form

$$\Delta E[L_{t+1}] = \mu \, \delta \, E[L_t](1 - E[L_t]). \tag{2.27}$$

(A similar result for the continuous-time model was established in [15].)

Equation (2.27) represents a difference analog of the differential equation $dx/dt = \beta \, x \, (1 - x)$, where $\beta = \mu \, \delta$. The latter's solution is *the logistic curve*, a classical concept in the theory of learning; for example, see a survey in [4]. The discrete form of the logistic curve is described by

$$E[L_t] = \frac{1}{1 + (\frac{1}{\lambda} - 1) \exp(-\beta t)}. \tag{2.28}$$

This function is monotonically increasing from $\lambda > 0$ (at $t = 0$) to 1 (as $t \to +\infty$).

As a rule, the solutions of similar difference and differential equations are not the functions of the same form; generally speaking, *the logistic curve* in the discrete form (2.28) is not a solution of (2.27). Therefore, we will establish conditions under which function (2.28) well approximates the solution of Eq. (2.27).

For the sake of local simplifications, introduce the compact notation $x_t = \frac{1}{1 + ba^t}$ for function (2.28).

First, prove that the difference equation describing (2.28) will turn into the differential equation $dx/dt = \beta \, x(1-x)$ as $\Delta t \to 0$. (Here Δt denotes the discrete time increment.)

In view of the compact notation, write $x_{t+\Delta t} = \frac{1}{1 + ba^{t+\Delta t}}$ and further transform this expression. The following chain of transformations are correct as $\Delta t \to 0$ (in fact, under the condition $\ln(a) \Delta t \ll 1$):

$$x_{t+\Delta t} = \frac{1}{1 + ba^{t+\Delta t}} \approx \frac{1}{1 + ba^t(1 + \ln(a)\Delta t)} = \frac{1}{1 + ba^t + ba^t \ln(a)\Delta t}$$

$$= \frac{1}{1 + ba^t} \frac{1}{1 + ba^t(1 + ba^t)^{-1} \ln(a)\Delta t}$$

$$\approx \frac{1}{1 + ba^t} \left(1 - ba^t(1 + ba^t)^{-1} \ln(a)\Delta t\right)$$

$$= \frac{1}{1 + ba^t} - \frac{ba^t}{(1 + ba^t)^2} \ln(a)\Delta t = x_t + x_t(1 - x_t)\ln(a)\Delta t.$$

Therefore, $x_{t+\Delta t} = x_t + x_t(1 - x_t)\ln(a)\Delta t$ if $\ln(a)\Delta t \ll 1$. As a result, $\frac{x_{t+\Delta t} - x_t}{\Delta t} = \ln(a)x_t(1 - x_t)$, which completes the proof.

Obviously, for $ln(a) < <1$ and $\Delta t = 1$ all these transformations remain in force, and the difference equation describing the logistic curve in the discrete form (2.28) is well approximated by (2.27).

In accordance with the intermediate notations, $ln(a) = \beta = \mu \delta = \mu/K$ and hence the condition $ln(a) \ll 1$ can be written as $\mu/K \ll 1$. Thus, for a large dimension K of the set of all states of the environment the logistic curve (2.28) is well approximated by the difference Eq. (2.27).

The logistic learning curve (2.28) is classical in the theory of learning [4]. At the same time, this curve is a special case of the learning-to-learn model that corresponds to the uniform distribution of a "large" set of all states of the environment and the proportional relation between the intensity of testing different states of the environment and the reached learning level.

If the learning curve is logistic (see 2.28), then the expected series length has the form $N_t = \lambda(1 - \lambda)^{-1}\exp(\beta t)$ generated by the compact difference equation $N_{t+1} = \exp(\beta)T_t$.

Hyperbolic learning curve. In another special case of auto-learning, the intensity of testing different states of the environment is decreasing in the learning level: $\Gamma_t = (1 - E[L_t])^a$, where $a > 0$. In practice, this relation well describes the limited cognitive and/or computational capabilities of a learned actor (in particular, the finite capacity of short-term memory).

We will derive a difference equation of the learning level in this case by analogy with the logistic learning curve for the equally probable states of the environment; see the previous paragraph. In this special case, expression (2.26) remains in force too. Substituting $\Gamma_t = (1 - E[L_t])^a$ into it gives

$$\Delta E[L_{t+1}] = \Gamma_t \delta(1 - E[L_t]) = \mu \delta(1 - E[L_t])^{1+a}. \tag{2.29}$$

Equation (2.29) represents a difference analog of the differential equation $dx/dt = \beta(1 - x)^{1+a}$, where $\beta = \mu\delta$. The latter's solution is *the hyperbolic learning curve*, a classical concept in the theory of learning; for example, see a survey in [4] and the pioneering papers [6, 7].

The discrete form of the hyperbolic curve is described by

$$E[L_t] = 1 - \frac{1}{(1 + a\beta t)^{1/a}}. \tag{2.30}$$

This function is monotonically increasing from 0 (at $t = 0$) to 1 (as $t \to +\infty$).

Like for the logistic curve, introduce the compact notation $x_t = 1 - \frac{1}{(1 + a\beta t)^{1/a}}$ and use the same considerations under the conditions $\beta \ll 1$ and $a\beta \ll 1$ to obtain $x_{t+1} = x_t + \beta(1 - x_t)^{a+1}$.

The condition $\beta \ll 1$ is equivalent to $\mu \delta = \mu/K \ll 1$. Consequently, the hyperbolic curve satisfies the difference Eq. (2.29) for a "large" dimension K of the set of all states of the environment.

In this case, the expected series length has the form $N_t = (1 + a\beta t)^{1/a} - 1$.

The difference equation of the expected series length is

$$N_{t+1} = \left[(N_t + 1)^a + a\beta\right]^{1/a} - 1.$$

In particular, for $a = 1$ the equation turns into $N_{t+1} = N_t + \beta$.

Thus, the hyperbolic learning curve (2.30) (its difference analog $E[L_t] = E[L_{t-1}] + \beta\left(1 - E[L_{t-1}]\right)^{1+a}$) is a special case of the learning-to-learn model that corresponds to the uniform distribution of a large set of all states of the environment and a decreasing relation between the intensity of testing different states of the environment and the reached learning level.

Auto-learning. Consider a continuous auto-learning model as follows. Let the dynamics of the learning level $z(t) \in [0, 1]$, $t \geq 0$, be described by the differential equation

$$\dot{z}(t) = \gamma (1 - z)\tilde{p}(z) \tag{2.31}$$

with an initial condition $z(0) = \lambda \in [0, 1)$, where $\gamma > 0$; $\tilde{p}(\cdot) : [0, 1] \rightarrow (0, A]$ is a continuous function; $0 < A < +\infty$. (If $\tilde{p}$ means probability, then $A = 1$.)

Due to the above assumptions, we have the following.

(a)   The solution of Eq. (2.31) exists and is unique.
(b)   The relation $z(t)$ is a strictly monotonically increasing function, i.e., $\forall t \geq 0$: $\dot{z}(t) \leq \gamma$.
(c)   If $z(0) = 0$, then $\forall t \geq 0$: $z(t) \leq 1 - \exp(-\gamma A t)$.
(d)   The relation $z(t)$ is slowly asymptotic, i.e., $\lim_{t \rightarrow +\infty} z(t) = 1$, $\lim_{t \rightarrow +\infty} \dot{z}(t) = 0$.

Different auto-learning curves can be obtained by varying $\tilde{p}(z)$. Special cases include many of the learning curves considered:

(1)   *the exponential curve* ("degenerate case"—auto-learning is replaced by standard learning, $\tilde{p}(z) \equiv 1$

$$\lambda = 0; \dot{z}(t) = \gamma(1 - z); z(t) = 1 - \exp(-\gamma t).$$

(2)   *the logistic curve*

$$\tilde{p}(z) = z, \lambda > 0; \dot{z}(t) = \gamma z(1 - z); z(t) = \frac{1}{1 + (\frac{1}{\lambda} - 1)\exp(-\gamma t)}.$$

(3) *the hyperbolic curve*

$$\tilde{p}(z) \equiv (1-z)^a, a > 0, \lambda = 0; \dot{z}(t) = \gamma(1-z)^{1+a} z(t) = 1 - \frac{1}{(1+a\gamma t)^{1/a}}.$$

The graphs of the three learning curves with $\gamma = 0.1$ and $a = 1$ are shown in Fig. 2.1.

Complex learning curves. The framework of the auto-learning model (2.31) can be used to construct "complex learning curves" [4], e.g., the curves with plateau. Figure 2.2 illustrates the graph of a learning curve determined by the equation $\dot{z}(t) = \gamma(1-z)\tilde{p}(z)$, where $\tilde{p}(z) = 1 + \frac{1}{2}\text{Sin}(6\,t)$, with the parameter $\gamma = 0.1$. The second term in the right-hand side may describe productivity variations during a working day (e.g., the warming-up or fatigue effects).

Forgetting effects. The general auto-learning model (2.31) may also reflect forgetting effects (though, under the assumption that the function $\tilde{p}(\cdot)$ may have negative values). Suppose the learning process was developing before a time T_0, and then the reached learning level started decreasing, e.g., in accordance with the law

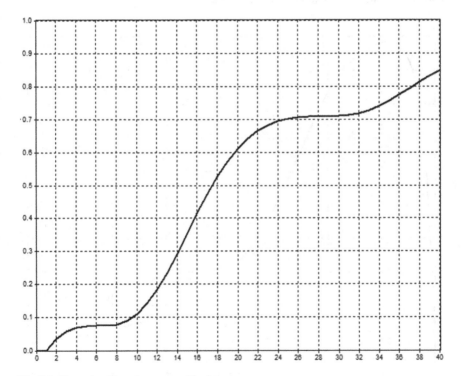

Fig. 2.2 Example of learning curve with plateau

$$\tilde{p}(z) = \begin{cases} 1, & t \le T_0, \\ -\frac{1}{2}z^2, & t > T_0. \end{cases}$$

The corresponding learning curve with the parameters $\gamma = 0.1$ and $T_0 = 40$ is shown in Fig. 2.3.

Some generalizations. The auto-learning model (2.31) allows for several extensions, namely, transitions to learning-by-doing and collective learning.

Consider *the learning-by-doing model* in which a learned actor (agent) can choose the intensity $w(t) \ge 0$ of his/her activity (e.g., the amount of work executed per unit time; the number of current states of the environment analyzed per unit time, etc.). The amount of executed work $W(t) = \int_0^t w(\tau)d\tau$ can be treated as the experience accumulated by the agent, his/her "productive internal time" [4, 16].

Replacing the function $\tilde{p}(z)$ with the intensity $w(t)$ in Eq. (2.31) yields the differential equation

$$\dot{z}(t) = \gamma(1 - z)w(t); \tag{2.32}$$

its solution is the "exponential" learning curve

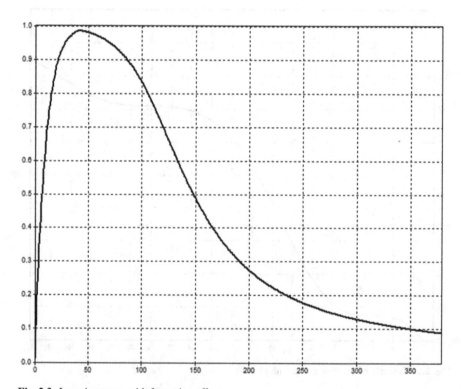

Fig. 2.3 Learning curve with forgetting effects

$$z_w(t) = 1 - \exp(-W(t)). \tag{2.33}$$

Following [16], assume function (2.33) specifies the probability of achieving the result at a time t—the share of all successful actions of the agent. Then the cumulative expected result can be calculated as

$$W_+(t) = \int_0^t z_w(\omega)w(\omega)d\omega = \int_0^t 1 - \exp\left(-\int_0^\omega w(\tau)d\tau\right)w(\omega)d\omega. \tag{2.34}$$

Let $T \geq 0$ be a given interval and also let W_0 be a maximum amount of work in accordance with the agent's capabilities. In view of (2.32)–(2.34), expected result maximization is the dynamic programming problem

$$W_+(T) \to \max_{w(\cdot),\,W(T)\leq W_0}.$$

Similar optimization problems (in particular, subject to the agent's cost constraints, etc.) interpreted in terms of the agent's optimal learning strategy were considered in [16]. Also see Chap. 3 of this book.

Concluding this section, we will describe the process of *collective learning* [16] in terms of auto-learning.

Up to this point, the agent's learning process has been considered under the assumption that the agent uses his/her "own" experience only. However, the members of real groups are exchanging their experience: an agent can gain experience by observing the activity of the others (their achievements or challenges). Such models were described in [4, 16]. For a proper reflection of this effect, let the experience $\tilde{p}$ accumulated by an agent be dependent on the learning levels of the other agents.

Consider n agents. Introduce the following notations: i as the agent's number; z_i as his/her learning level; $z = (z_1, z_2, \ldots, z_n)$ as the learning levels vector. For each agent, write an analog of Eq. (2.31):

$$\dot{z}_i(t) = \gamma_i(1 - z_i)\tilde{p}_i(\mathbf{z}), \quad i = \overline{1, n}. \tag{2.35}$$

In model (2.35), agents may have different influence on each other as follows.

- If $\frac{\partial \tilde{p}_i(\mathbf{z})}{\partial z_j} > 0$, then agent i adopts experience from agent j.
- If $\frac{\partial \tilde{p}_i(\mathbf{z})}{\partial z_j} < 0$, then the experience of agent j "confuses" agent i.
- If $\frac{\partial \tilde{p}_i(\mathbf{z})}{\partial z_j} \equiv 0$, then the experience gained by agent j does not affect agent i.

Within the framework of model (2.35), the optimal collective learning problem for a group of agents can be formulated and solved in the same way as in [16].

Thus, in Chap. 2 the design and adoption models of CA technology have been considered. The main results include the following.

- *The basic learning curve* for technology design (Formula 2.1) has been obtained and its properties (Propositions 1 and 2) have been established.
- The learning curve has been approximated in a series of important special cases (Formulas 2.3 and 2.6).
- The *expected time* of reaching a required learning level has been estimated (Formula 2.7) and its properties have been established (Proposition 3).
- Also:

  - the *integration models* of partial technology components (Formulas 2.15, 2.17, 2.19, 2.21, 2.25, 2.26 and 2.16, 2.18, 2.20 for the learning curves and expected times, respectively) and
  - the *models of learning to learn* and *auto-learning*.

  have been suggested and analyzed.

Note that the exponential (2.2), logistic (2.27) and hyperbolic (2.30) curves (the classical ones in the theory of learning [4]) have turned out to be special cases of the suggested auto-learning model.

For the classical learning curves, *the expected series length* (an index that has been introduced in Chap. 2) satisfies linear difference equations as follows:

- for the exponential learning curve, $N_{t+1} = \exp(\gamma)N_t + (\exp(\gamma) - 1)$;
- for the logistic learning curve, $N_{t+1} = \exp(\beta)N_t$;
- for the hyperbolic learning curve of first degree, $N_{t+1} = N_t + \beta$.

Perhaps, this fact deserves further study.

A promising line of research is to formulate and solve *management problems* for the design and development of technologies based on their models presented in Chap. 2. Some of such management problems will be considered in Chap. 3.

References

1. Belov M, Novikov D (2018) Methodology of complex activity. Lenand, Moscow, 320 pp (in Russian)
2. Business Process Model and Notation (BPMN), v2.0.2. http://www.omg.org/spec/BPMN/2.0
3. Novikov D (2013) Theory of control in organizations. Nova Science Publishers, New York, 341 pp
4. Novikov D (1998) Laws of iterative learning. Trapeznikov Institute of Control Sciences RAS, Moscow, 98 pp (in Russian)
5. Ebbinghaus H (1885) Über das Gedächtnis. Dunker, Leipzig, 168 pp
6. Thurstone L (1919) The learning curve equation. Psychol Monogr 26(3):1–51
7. Thurstone L (1930) The learning function. J Gen Psychol 3:469–493
8. Tolman E (1934) Theories of learning. In: Moss FA (ed) Comparative psychology. Prentice Hall, New York, pp 232–254
9. Atkinson R, Bower G, Crothers J (1967) Introduction to mathematical learning theory. Wiley, New York, 429 pp
10. Bush R, Mosteller F (1955) Stochastic models for learning. Wiley, New York, 365 pp
11. Hull C (1943) Principles of behavior and introduction to behavior theory. D. Appleton Century Company, New York, 422 pp

12. Wright T (1936) Factors affecting the cost of airplanes. J Aeronaut Sci 3(4):122–128
13. Crawford J (1944) Learning curve, ship curve, ratios, related data. Lockheed Aircraft Corporation, pp. 122–128
14. Henderson B (1984) The application and misapplication of the learning curve. J Bus Strategy 4:3–9
15. Leibowitz N, Baum B, Enden G, Karniel A (2010) The exponential learning equation as a function of successful trials results in sigmoid performance. J Math Psychol 54:338–340
16. Novikov D (2012) Collective learning-by-doing. IFAC Proc Vol 45(11):408–412

Chapter 3
Models of Technology Management

In Chap. 2, the basic design and adoption model for the technology of *complex activity* (CA) has been presented [1]. In the current chapter, a set of management problems arising in the design and adoption of the new technologies of complex activity will be considered, which includes the following problems: optimal learning (optimal choice of typical solutions); resource allocation in technological networks; optimal strategy development for the transition from technology design to productive use.

Let us summarize the key results of the previous chapters. The execution of different types of CA is described by a discrete process as follows: at each time, precisely one CA element is implemented under the assumption that the *environment* takes precisely one value from the set of all possible states of the environment. If at some time the environment evolves to a new state never observed before, then an event of *uncertainty* occurs. This event leads to additional cost for creating or adapting the technology to the new conditions. If the environment returns in this state at one of the subsequent times, then no additional cost is required.

Let the SPSE be composed of K different values. Assume at each time the environment takes precisely one of them regardless of the past states. Denote by $p_k > 0$ the probability that the environment takes the kth value (obviously, $\sum_{k=1}^{K} p_k = 1$).

Within the framework of this model, the implementation process of different phases in *the technology's life cycle* is completely characterized by the dynamics of the current states of the environment: which values from the SPSE the environment has already taken (and how many times) and which has not; see the details in Chap. 2. *The maturity level of the technology* (an analog of the learning level) is used here. Recall that the sequence of its values is called *the learning curve*. The index L_t gives the share of the states of the environment for which the technology has been tested or adapted during the past t times, or the probability that at the next time $(t + 1)$ the environment will take one of its previous values:

$$L_t = 1 - \sum_{k=1}^{K} p_k (1 - p_k)^t. \tag{3.1}$$

© Springer Nature Switzerland AG 2020
M. V. Belov and D. A. Novikov, *Models of Technologies*, Lecture Notes
in Networks and Systems 86, https://doi.org/10.1007/978-3-030-31084-4_3

While sequence (3.1) represents the learning curve, the sequence

$$Q_t = 1-L_t = \sum_{k=1}^{K} p_k(1 - p_k)^t \qquad (3.2)$$

can be interpreted as *the error curve* (the probability that at the next time $(t + 1)$ the environment will take one of its new values not observed before).

Note that the technology adoption process should be treated as *the actor's learning process*; see the classical works [2–7] and also the modern learning models in the surveys [8–11], in which the learning curves (3.1) are typical.

In this chapter, model (3.1) and (3.2) will be used to formulate and solve the following *management problems*: the optimal learning problem—find a partition of the set of all possible states of the environment into a finite number of subsets that minimizes the expected error (Sect. 3.1) or the entropy (Sect. 3.2); the optimal resource allocation problem in technological networks (Sect. 3.3); the optimal transition problem from technology design to productive use (Sect. 3.4).

3.1 Optimal Learning: Typical Solutions

Consider an actor (*agent*) that makes certain *decisions* during his/her activity. Let *the efficiency* $x \in [0, 1]$ of the agent's decisions be described by a function $f(x, \theta)$ that depends on the realized *state of the environment* $\theta \in [0, 1]$. For the sake of simplicity, assume $\arg\max_{x\in[0,1]} f(x, \theta) = \theta$. An example of such a function is $f(x, \theta) = 1-(x - \theta)^2$.

Assume the agent distinguishes among K values of the state of the environment that are realized with probabilities $\{p_k\}$, $k = \overline{1, K}$. Partition the unit interval into K sequential subintervals Δ_k of the lengths $\{p_k\}$ with the limits $\left[\sum_{i=0}^{k-1} p_i, \sum_{i=0}^{k} p_i\right]$, where $p_0 = 0$.

Consider *a discrete learning process* of the following form. At each time, a certain state of the environment is realized; if some state of the environment is realized again, then the agent chooses the corresponding optimal decision $x^*(\theta)$, where $x^*(\theta) = \arg\max_{x\in[0,1]} f(x, \theta)$; if some state of the environment (e.g., j-th) is newly realized (never observed before), then the agent chooses an arbitrary decision from the corresponding subinterval Δ_j. On the one hand, this decision principle formally matches the Aumann's model [12], in which decision-making based on act-rationality and rule-rationality was considered. On the other hand, the model suggested in this section well reflects the ideology *typical solutions* (see Chap. 4 of the book), which is widespread in situational and adaptive management.

Let the function $f(\cdot, \theta)$ be uniformly l-Lipschitzian, where $l > 0$, for any states of the environment. (If $l \leq 0$, the efficiency turns out to be independent of the decisions.) Then the maximum expected error of the agent's decision at time t (the difference between the efficiencies of the chosen and optimal solutions) can be estimated as $\sum_{k=1}^{K} p_k(1 - p_k)^t l \ p_k$. Also see Formula (3.2).

Fixing an arbitrary integer $K \geq 1$ and a minimum *threshold* ρ, $0 < \rho \leq \frac{1}{K}$, for distinguishing the states of the environment, we may formulate the optimal partition problem of all possible states of the environment (the unit interval) into K subsets as follows:

$$Q(\{p_k\}, t) = \sum_{k=1}^{K} (p_k)^2 (1 - p_k)^t l \ \rightarrow \min_{\{p_k \geq \rho\}: \sum_{k=1}^{K} p_k = 1} . \qquad (3.3)$$

Note that the nonzero threshold ρ allows us to avoid the trivial solution $p_1 = 1$ and $p_j = 0$, where $j = \overline{2, K}$.

For *the uniform distribution* $(p_k = 1/K)$, the objective function of problem (3.3) takes the form

$$Q_0(K, t) = \frac{l}{K}(1 - \frac{1}{K})^t. \qquad (3.4)$$

Problem (3.3) can be interpreted as seeking for an optimal set of typical solutions that minimize the expected error of the current decisions at a given time.

Proposition 1 $\forall \rho \in (0, \ 1/K] \ \exists t(\rho)$ *such that* $\forall \tau > t(\rho)$ *the unique solution of the problem*

$$Q(\{p_k\}, \tau) \rightarrow \min_{\{p_k \geq \rho\}: \sum_{k=1}^{K} p_k = 1} \qquad (3.5)$$

is the uniform distribution.

The proof of Proposition 1 will rest on an intermediate result as follows.

Lemma 1 $\forall \rho \in (0, \ 1/K] \ \exists t(\rho)$ *such that* $\forall \tau > t(\rho)$ *the function* $Q(\{p_k\}, \tau)$ *is strictly convex in the variables* $\{p_k\}$.

The proof of Lemma 1 Fix an arbitrary number $k = \overline{1, K}$. Omitting the subscript k below, demonstrate that $\forall \rho \in (0, \ 1/K] \ \exists t(\rho)$ such that $\forall \tau \geq t(\rho)$ the function $G(p) = p^2 (1 - p)^t$ is convex in p. Calculate the second-order derivative of the function $G(\cdot)$:

$$\frac{d^2 G(p)}{dp^2} = (1-p)^{t-2} \big[2(1-p)^2 - 4 \, p \, t(1-p) + p^2 t(t-1) \big]. \qquad (3.6)$$

As $t(\rho)$ choose the maximum root of the quadratic equation

$$\forall p \in [\rho, \ 1 - \rho] \, 2(1 - p)^2 - 4 \, p \, t(1 - p) + p^2 t(t - 1) = 0, \text{ where } t \geq 2. \quad (3.7)$$

Equation (3.7) has a nonnegative solution, since its coefficient at the highest (quadratic) term is strictly positive. Clearly, any $\tau \geq t(\rho)$ satisfies the system of inequalities $\frac{d^2 G(p)}{dp^2} > 0$, $\forall p \in [\rho, 1 - \rho]$. Hence, due to the continuity of the right-hand side of expression (3.6) in t, it follows that $\forall p \in [\rho, 1 - \rho], \forall \tau > t(\rho)$: $\frac{d^2 G(p)}{dp^2} > 0$.

Thus, each term of $\sum_{k=1}^{K} (p_k)^2 (1 - p_k)^t l$ represents a convex function of p_k because the Lipschitz constant is nonnegative by definition. Consequently, their sum is also a convex function, which concludes the proof of Lemma 1.

Now, prove the main result—Proposition 1 in Chap. 2. Fix an arbitrary time $t > 0$. Assume $\{q_k\}$ is the solution of problem (3.5) for $t \geq t(\rho)$ and, in addition, there exists a pair $i, j \in \overline{1, K}$, such that $i \neq j$ and $q_i \neq q_j$. For the sake of definiteness, let $j > i$. Due to the strong convexity of the objective function, we have

$$Q(\{q_k\}, t) > Q(q_1, \ldots, q_{i-1}, \frac{q_i + q_j}{2}, q_{i+1}, \ldots, q_{j-1}, \frac{q_i + q_j}{2}, q_{j+1}, \ldots, q_K, t),$$

which contradicts the assumption. This means that in the optimal solution all values $\{q_k\}$ are the same. The uniqueness of this optimal solution follows from the strong convexity of the objective function. The proof of Proposition 1 is complete.   □

Interestingly, the solution of problem (3.5) does not depend on the Lipschitz constant l.

Example 1 Choose $K = 2$. The graph of $Q(p, t)$ is shown in Fig. 3.1.

Proposition 1 can be generalized to the following case. Denote by $C_k(p_k)$ the agent's "losses" incurred by the first realization of the kth state of the environment (in a practical interpretation, *the cost of* obtaining the optimal solution in this situation). This cost can be determined as the computational complexity of the optimal solution in this situation. As it will be established in Chap. 4, the corresponding computational cost has linear or even superlinear growth with respect to the "size" of the optimization domain.

Fig. 3.1 Graph of Q(p, t) in Example 1

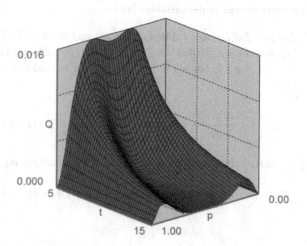

The optimal partition problem of the set of all possible states of the environment (the unit interval) into K subsets in terms of the minimum expected losses at the current time t takes the form

$$Q_C(\{p_k\}, t) = \sum_{k=1}^{K} C_k(p_k) p_k (1 - p_k)^t \rightarrow \min_{\{p_k \geq \rho\}: \sum_{k=1}^{K} p_k = 1} . \tag{3.8}$$

Corollary 1 *Let the functions* $C_k(\cdot), k = \overline{1, K},$ *be strictly positive and have bounded first- and second-order derivatives. Then* $\forall \rho \in (0, \ 1/K] \ \exists t(\rho)$ *such that* $\forall \tau > t(\rho)$ *the unique solution of problem* (3.8) *is the uniform partition (uniform probability distribution).*

This corollary is proved by analogy with Proposition 1, with the only exception that the function $G(p)$ is replaced by the function $G_C(p) = C(p)p(1-p)^t$. Calculate the second-order derivative of $G_C(\cdot)$:

$$\frac{d^2 G_C(p)}{dp^2} = (1-p)^{t-2}[C''(p)p(1-p)^2$$
$$+ 2C'(p)(1-p)(1-p-pt) + C(p)pt(t-1)]. \tag{3.9}$$

By the hypotheses of Corollary 1 the coefficient at the highest (quadratic) term in the right-hand side of expression (3.9) is strictly positive while the other coefficients are bounded. The proof of Corollary 1 is complete. $\qquad\square$

Next, assume the agent will gain some *payoff* $H_k(p_k)$ if the kth state of the environment is realized one or more times again. In this case, the optimal partition problem of the set of all possible states of the environment (the unit interval) into K subsets in terms of the maximum expected utility (the difference between the payoff and cost) at time t can be written as

$$Q_{H,C}(\{p_k\}, t) = \sum_{k=1}^{K} p_k \{[1 - (1 - p_k)^t] H_k(p_k) - C_k(p_k)(1 - p_k)^t\} \rightarrow \max_{\{p_k \geq \rho\}: \sum_{k=1}^{K} p_k = 1} .$$
$$\tag{3.10}$$

Theorem 1 (on optimal typical solutions) *Let the functions* $H_k(\cdot), k = \overline{1, K},$ *be such that the functions* $x H_k(x)$ *are strictly concave for* $x \in [0, \ 1],$ *and also let the functions* $C_k(\cdot), k = \overline{1, K},$ *satisfy the hypotheses of Corollary 1. Then* $\forall \rho \in (0, \ 1/K] \ \exists t(\rho)$ *such that* $\forall \tau > t(\rho)$ *the unique solution of problem* (3.10) *is the uniform partition.*

The proof of Theorem 1 By the hypotheses of this theorem and Corollary 1, each of the terms in the objective function (3.10) is a strictly concave function as the difference of strictly concave and strictly convex functions. Hence, the function

$(\{p_k\}, t)$ is concave in $\{p_k\}$. Using the same considerations as in the proof of Proposition 1, we can easily show that the optimal values of $\{p_k\}$ are the same. The proof of Theorem 1 is complete. $\square$

Proposition 1 claims that, for any threshold, there exists a time since which the uniform probability distribution will minimize the expected error of the agent's decisions. A natural question, which actually characterizes the converse property, is as follows. For sufficiently large times, does there exist a threshold under which the uniform distribution is optimal? The answer is affirmative.

Proposition 2 $\forall t \geq [2K - 3/2 + \sqrt{2(K^2 - K - 1)}] \; \exists \rho(t) \leq 1/K$ *such that one of the solutions of the problem*

$$Q(\{p_k\}, t) \to \min_{\{p_k \geq \rho(t)\} : \sum_{k=1}^{K} p_k = 1} \tag{3.11}$$

is the uniform partition.

The proof of this result seems trivial: under the hypotheses of Proposition 2, the convexity condition

$$2(1-K)^2 - 4t(1-1/K)/K + t(t-1)/K^2 \geq 0$$

holds for $\rho(t) = 1/K$. Also see expressions (3.6), (3.7) and (3.9).

Up to this point, the number K of pairwise distinguishable states of the environment has been assumed to be fixed. Now, consider how this number affects the expected error, i.e., find the optimal value of K. In view of Proposition 1 and Theorem 1, it suffices to study the class of uniform distributions only. Direct analysis of Formula (3.4) gives the following result.

Proposition 3 *For any $t \geq 0$, there exists the unique "worst" value $K_*(t) = t + 1$ that maximizes the error.*

Example 2 The graph of (3.4) with $t = 50$ is shown in Fig. 3.2; here $K_* = 51$.

Error (3.4) achieves minimum under small or sufficiently large values of K. Hence, additional criteria should be introduced, e.g., the boundedness of the agent's cognitive capabilities, the relation between the learning level and the number of states of the environment, etc.

Indeed, up to this point the expected error (the objective function in the optimization problem (3.3)) has been adopted as the criterion. Now, for this role choose the learning level—the probability that a known state of the environment is realized; see (3.1).

Example 3 For the uniform distribution, the relation between the expected learning level and K has the form

$$L(K, t) = 1 - (1 - \frac{1}{K})^t. \tag{3.12}$$

Q(K,t=50)

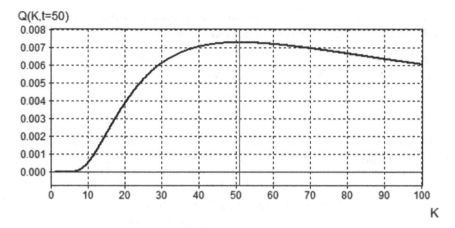

Fig. 3.2 Graph of Q(K, t = 50) in Example 2

The graph of (3.12) is presented in Fig. 3.3.
Consider the expected learning level maximization problem

$$L(\{p_k\}, t) = 1 - \sum_{k=1}^{K} p_k(1 - p_k)^\tau \to \max_{\{p_k \geq \rho\}: \sum_{k=1}^{K} p_k = 1} . \qquad (3.13)$$

Fig. 3.3 Graph of L(K, t) in
Example 3

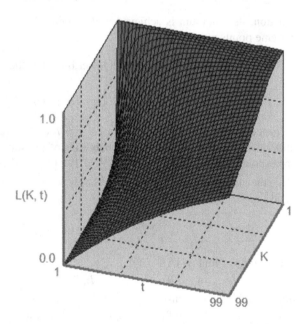

For this problem, an analog of Proposition 1 can be established using the convexity conditions of the objective function's terms as follows.

Proposition 4a $\forall \rho \in (0, 1/K] \exists t(\rho) = \frac{2}{\rho} - 1$ such that $\forall \tau > t(\rho)$ the unique solution of problem (3.13) is the uniform partition.

Which probability distribution is "worst" in terms of criterion (3.13)? The next result gives the answer.

Proposition 4b $\forall \rho \in (0, 1/K] \exists t^*(\rho) = \frac{2}{\rho} - 1$ such that $\forall \tau > t^*(\rho)$ the solution p^{min} of the problem

$$L_\tau = 1 - \sum_{k=1}^{K} p_k (1 - p_k)^\tau \to \min_{\{p_k \geq \rho\}: \sum_{k=1}^{K} p_k = 1}$$

has the form $p_{ii}^{min} = 1 - (K-1)\rho$, $p_{ik}^{min} = \rho, k \neq i, i = \overline{1, K}$.

Proof As it has been established earlier (see Proposition 1), $\forall \rho \in (0, 1/K] \exists t^*(\rho)$ such that $\forall \tau > t^*(\rho)$ the function $\sum_{k=1}^{K} p_k (1 - p_k)^\tau$ is strictly convex in $\{p_k\}_{k=\overline{1,K}}$. (The estimate $t^*(\rho) = \frac{2}{\rho} - 1$ follows directly from the positivity of the second-order derivative.) On a bounded convex set, a strictly convex function achieves maximum at one of its extreme points. The point p^{min} is an extreme point of the convex polyhedron $\{p_k \geq \rho, k = \overline{1, K}; \sum_{k=1}^{K} p_k = 1\}$. Due to the obvious symmetry of the objective function, its minimum is achieved at this point also (note that the values at all K extreme points are the same). □

Using Proposition 4b, calculate a lower bound of the value L_τ for $\tau > t^*(\rho)$:

$$L_\tau^{min}(\rho) = 1 - (K - 1)^\tau \rho^\tau [1 - (K - 1)\rho] - (K - 1)\rho(1 - \rho)^\tau.$$

In accordance with Proposition 4a, for $\tau > t^*(\rho)$ the value L_τ achieves maximum under the uniformly distributed probabilities of the states of the environment. The corresponding upper estimate is $L_\tau^{max} = 1 - (1 - \frac{1}{K})^\tau$ (interestingly, for $0 \leq \tau \ll t^*(\rho)$ the inequality $L_\tau^{min}(\rho) \geq L_\tau^{max}$ holds; see Fig. 1.1). Standard transformations yield

$$L_\tau^{max} = 1 - \exp(-\gamma(K)\tau),$$

where $\gamma(K) = \ln(1 + 1/(K - 1))$.

Since $\forall \rho \in (0, 1/K]$ $L_\tau^{min}(\rho) \leq L_\tau \leq L_\tau^{max}$ and $\lim_{\tau \to +\infty} L_\tau^{min}(\rho) = \lim_{\tau \to +\infty} L_\tau^{max} = 1$, then $\lim_{\tau \to +\infty} L_\tau = 1$ and $\lim_{\tau \to +\infty} (L_\tau^{max} - L_\tau^{min}(\rho)) = 0$. In fact, the following result has been argued.

Theorem 2 $\forall \rho \in (0, \ 1/K]$ *and* $\forall \tau > t^*(\rho)$,

(1) $L_{\tau}^{\min}(\rho)$ is increasing in ρ;

(2) $L_{\tau}^{\min}(\rho) \le L_{\tau}^{\max}$;

(3) $L_{\tau}^{\min}(\frac{1}{K}) = L_{\tau}^{\max}$;

(4) $\lim_{\tau \to +\infty}(L_{\tau}^{\max} - L_{\tau}^{\min}(\rho)) = 0$.

For any fixed time, the learning level is decreasing in the number K; see expression (3.12) and also Fig. 3.3. Moreover, by Proposition 3 the relation between the error and this number has a maximum point. Why cannot we choose $K = 1$, assuming that the set of all possible states of the environment is a singleton? This question seems natural but such an assumption will make the system's behavior independent of the states of the environment. Hence, a reasonable approach is to hypothesize the existence of K_0 <u>fundamentally different</u> states of the environment that require qualitatively different responses from the agent. (The number K_0 has to be a priori known.) On the one hand, this number can be determined from some objective laws or retrospective data (in the case of measurable uncertainty in the states of the environment), or determined using some heuristics/expertise (in the case of true uncertainty in the states of the environment). On the other hand, this number imposes an explicit lower bound on the number of different states of the environment (the inequality $K \ge K_0$) and must agree with the threshold ρ (the inequality $\rho \le \frac{1}{K_0}$) (Fig. 3.4).

Now, analyze which factors may restrict an infinite increase of the parameter K. The natural restrictions on the number K are as follows.

- The inequality $p_k \ge \rho$ implies $K \le 1/\rho$.
- Proposition 1 gives $t(\rho) \ge 2K - 1$.
- If δ is the agent's "differential threshold" for the values of the objective function, then $K \le 1/\delta$.

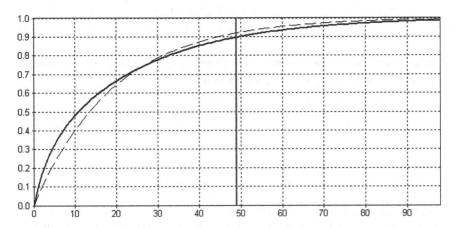

Fig. 3.4 Graphs of $L_{\tau}^{\min}(\rho)$ (solid line) and $L_{\tau}^{\max}$ (dashed line) for K = 20 and ρ = 0.04. Vertical line is $t^*(\rho) = 49$

Thus, a rational choice is to partition the set of all possible states of the environment into K equally probable "situations" so that (a) $K \geq K_0$, (b) K guarantees a reasonable compromise between the expected error and the learning level at a current time and (c) K satisfies the above-mentioned upper bounds. For each situation, the agent will find optimal or typical solutions during technology design.

Consider an example of the optimal management problem as follows. Given some K_0, l and ρ, the problem is to reach a required learning level L_{req} by a time τ so that the expected error will not exceed ε. Within Proposition 1, this system of requirements is consistent if there exists a positive integer K such that

$$\begin{cases} K_0 \leq K \leq \frac{1}{\rho}, \\ 1 - (1 - \frac{1}{K})^\tau \geq L_{\text{req}}, \\ \frac{l}{K}(1 - \frac{1}{K})^\tau \leq \varepsilon. \end{cases} \tag{3.14}$$

(Also see Formulas (3.4) and (3.12).) In accordance with Proposition 3, for $K_0 \leq K * (\tau)$ it suffices to check all inequalities (3.14) for the number $K = K_0$ (the greater numbers of K are pointless because they will simultaneously reduce the learning level and increase the expected error); for $K_0 > K * (\tau)$, the admissible values of the parameter K have to be found.

3.2 Entropy

Assume one of the two events may occur at each time—the realizations of the known or unknown (new) states of the environment. The former event has the probability $L(\{p_k\}, t)$ defined by (3.13). In the case of two possible events, *the entropy is*

$$S(t, \{p_k\}) = -L(\{p_k\}, t) \ln(L(\{p_k\}, t)) \\ -(1 - L(\{p_k\}, t)) \ln(1 - L(\{p_k\}, t)). \tag{3.15}$$

Study the relation between entropy (3.15) and $\{p_k\}$, K and t. More specifically, consider the entropy minimization problem at time t:

$$S(t, \{p_k\}) \to \min_{\{p_k \geq \rho\}: \sum_{k=1}^{K} p_k = 1} . \tag{3.16}$$

Theorem 3 (on entropy) $\forall \rho \in (0, \ 1/K] \ \exists \ t(\rho) = \frac{2}{\rho} - 1$ *such that* $\forall \tau > t(\rho)$ *the unique solution of problem (3.16) is the uniform partition.*

Theorem 3 directly follows from the fact that entropy (3.15) is minimum if one of the probabilities $L(\{p_k\}, t)$ or $(1 - L(\{p_k\}, t))$ achieves maximum. By Proposition 4 precisely the uniform distribution maximizes (3.13).

The result of Theorem 3 is not trivial: the maximum variety of the initial states (the uniform probability distribution of all possible states of the environment) not only minimizes the expected error and maximizes the learning level (Propositions 1 and 4, respectively) but also minimizes the entropy of the agent's learning states.

Proposition 5 *The maximum value of entropy* (3.15) *does not depend on the distribution* $\{p_k\}$, *being equal to* $\ln(2)$.

Indeed, Proposition 5 follows from the fact that expression (3.15) is maximized with respect to the time variable if $L(\{p_k\}, t) = 1 - L(\{p_k\}, t)$, i.e., if the known and new states of the environment are realized with the same probability.

For the uniform distribution, relation (3.15) between the entropy, time and parameter K takes the form

$$S(t, K) = \ln \left[\frac{\left(\left(\frac{K-1}{K}\right)^t - 1 \right)^{\left(\frac{K-1}{K}\right)^t}}{1 - \left(\frac{K-1}{K}\right)^t} \right]. \tag{3.17}$$

Entropy (3.17) achieves maximum at the time

$$t_S(K) = -\frac{\ln(2)}{\ln(1 - 1/K)}. \tag{3.18}$$

Note that $t_S(K) \leq t(\rho)$, i.e., the uniform distribution is optimal for the times considerably exceeding the characteristic time for achieving the maximum entropy (Figs. 3.5 and 3.6).

This model satisfies *the principle of determinism destruction* [11, 13] because there exists a maximum point of the entropy. (At the initial time, the entropy is 0,

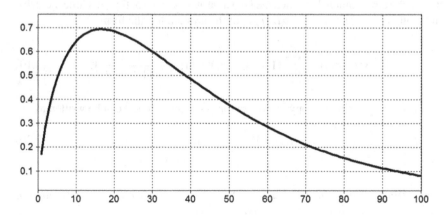

Fig. 3.5 Graph of S(K, t) under uniform distribution with K = 25 ($t_S(K) \approx 17$)

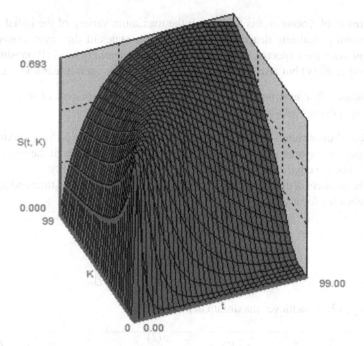

Fig. 3.6 Graph of S(K, t) under uniform distribution

meaning that the system is completely deterministic; hence, any state of the environment realized at the initial time will be new for the agent; the entropy vanishes as $t \to 0$.)

Now, consider the system with 2^K possible states. At time t, its dynamics are described by a K-dimensional binary vector, in which the kth component is 1 if the state of the environment has taken the kth value at least once before this time inclusive and 0 otherwise. The entropy $s(t, \{p_k\})$ of this system has the form

$$s(t, \{p_k\}) = \sum_{k=1}^{K} (1 - (1 - p_k)^t) \ln\left(\frac{1}{1 - (1 - p_k)^t}\right). \qquad (3.19)$$

The corresponding entropy minimization problem at time t can be written as

$$s(t, \{p_k\}) \to \min_{\{p_k \geq \rho\}: \sum_{k=1}^{K} p_k = 1}. \qquad (3.20)$$

Proposition 6 $\forall \rho \in (0, \ 1/K] \ \exists \ t(\rho)$ *such that* $\forall \tau > t(\rho)$ *the unique solution of problem* (3.20) *is the uniform partition.*

This result is established by analogy with Proposition 1 and Theorem 1. (For sufficiently large t, we should demonstrate the strict convexity of the function $(1 - (1-p)^t)\ln\left(\frac{1}{1-(1-p)^t}\right)$ in the variable p.) The technical details are therefore omitted.

3.3 Technological Networks

Recall that *the technology of CA* has been defined as a system of conditions, criteria, forms, methods and means for achieving a desired goal. A sequence of actions (the logical, temporal and process structures of CA [14], i.e., *technological networks*) is traditionally described using *graph theory*. This language provides the structural description (the connections between the whole and its parts, the connections between different elements), the cause-effect description and the functional description (the system's behavior and interaction with the environment, etc.). Really, network models well reflect the cause-effect relations between the CA elements: the descriptive and predictive function (from the causes to their effects), the explanatory function (from the effects to their causes) and the normative function (from the causes to the optimal effects or from the optimal causes to the required effects).

There are different classes of models that characterize the significant properties of a technological structure as follows:

- the information-logical models of science and technologies;
- the semantic, logical and Bayesian networks, namely, probabilistic logic networks (PLNs) [15], Markov logic networks [16] and binary neural networks [17];
- the knowledge representation models: the production models, the networked models (semantic networks, ontologies), the frame models, and others; see surveys in [18, 19];
- the models of science development in terms of bibliometry and citation networks [20];
- the knowledge epidemics models for ideas diffusion [21].

In addition, using the concepts of the technologies readiness level (TRL) and integration readiness level (IRL), recently many researchers have considered the maximization problems of the systems readiness level (SRL) subject to different TRL and IRL constraints [22, 23].

However, all these classes of models still need to be adapted for the management problems of CA technology design and adoption. Let us analyze the capabilities of the network active system models for these purposes.

A *network active system* (NAS) is defined as follows. Consider a *network* $G = (N, E)$ (a directed and connected graph without cycles), where $N = \{1, 2, \ldots, n\}$, $n \geq 2$, is a finite set of nodes (the *agents* implementing elementary technologies) and $E \subseteq N \times N$ is a finite set of edges (the logical relations between them). The nodes of this graph have *proper* numbering, which reflects the prior beliefs about the cause-effect relations between the activities of different agents.

The entire network can be treated as the model of some complex technology while its subgraphs as the models of partial technologies.

Denote by $L_i = \{j \in N | (j; i) \in E\}$ the set of the immediate *predecessors* of agent i in the network G and by $R_i = \{j \in N | (i; j) \in E\}$ the set of the immediate *followers* of agent i, $i \in N$. Designate as G_N the set of all possible networks with correct numbering that connect all nodes from the set N.

Assume the network has a unique *output*—node n. (An output is a node without any outgoing edges.) Let $M_0 \subseteq N$ be the set of all *inputs* of this network. (An input is a node without any incoming edges.) Let M_k be the set of all nodes with the incoming edges only from the nodes belonging to the sets $\{M_j\}$, where $j = \overline{0, k-1}, k = \overline{1, m}$, $m \leq n - 1$ and $M_m = \{n\}$. (The value $k(i)$ is *the rank of* node i belonging to the set 'M_k'.) Obviously, the output's rank $k(n)$ is the length of the maximal path from the network inputs to its output.) The collection of all $\{M_k\}$, $k = \overline{0, m}$, is a partition of the set N.

Introduce the notation $M^k = \cup_{j=0}^{k-1} M_k$, $k = \overline{1, m}$, and also let $M^0 = \emptyset$. Designate as $S_0 = \{n\}$, $S_k \subseteq N$, the set of all nodes of the graph G with the outgoing edges only to the nodes belonging to the set S_{k-1}, $k = 1, 2, \ldots, k(n)$. Since the graph G is connected, $\cup_{k=0}^{k(n)} S_k = N$. Denote by W_i the set of all predecessors of agent i, i.e., the nodes that have a path to node i. Since the graph G is connected, $W_n = N/\{n\}$. The rank of a node can be interpreted as the maturity level of a corresponding technology.

Let the NAS be *binary* in the sense that agent i chooses a binary *action* $y_i \in \{0; 1\}$, which gives a binary *result* $z_i \in \{0; 1\}$ of his/her activity. (For example, "0" means "the action is not implemented" or "the result is not achieved," while "1" means "the action is implemented" or "the result is achieved".) Designate as y_D and z_D, where $D \subseteq N$, the action and result vectors of all agents, respectively.

The agent's result depends on his/her actions and also on the results of other agents used by him/her. This dependence will be described by the logical *technological function* $Q_i : \{0; 1\}^{|N_i|} \to \{0; 1\}$, i.e., $z_i = y_i Q_i(z_{N_i})$. For $i \in M_0$, $N_i = \emptyset$; hence, let $z_i = y_i Q_i(z_0)$, where z_0 is the l-dimensional vector of all *inputs* of the network ($l = |M_0|$). Assume the choice $y_i = 1$ incurs some cost $c_i \geq 0$ on agent i.

The simplest examples of technological functions are *the conjunctive function* $Q_i^{\min}(z_{N_i}) = \min_{j \in N_i} \{z_j\}$ (agent i obtains the desired result only if all his/her immediate predecessors have achieved their own results) and *the disjunctive function* $Q_i^{\max}(z_{N_i}) = \max_{j \in N_i} \{z_j\}$ (agent i obtains the desired result if at least one of his/her immediate predecessors has achieved his/her own result).

Consider an actor that manages the NAS (hereinafter, such an actor will be called *the Principal*). If the Principal knows the graph G, the technological functions $\{Q_i(\cdot)\}$ and also the cost functions $\{c_i\}$ of all agents, then he/she may implement the following algorithm for each node i of the graph G.

– Find the function $Q^i(y_{W_i})$ determining the relation between the result z_i of agent i and the actions vector y_{W_i} of all his/her predecessors. (This function can be treated as *the aggregate technology of agent i*; for agent n, this is the aggregate technology of the entire NAS.)

- Find the set

$$A_i = \{(y_{W_i}) \in \{0, 1\}^{|W_i|} \,|\, Q^i (y_{W_i}) = 1\} \tag{3.21}$$

of the actions vectors for achieving the result of agent i.
- Find the set

$$A_i^* = \mathrm{Arg} \min_{(y_{W_i}) \in A_i} \sum_{j \in W_i} c_j \tag{3.22}$$

of the actions vectors for achieving the result of agent i with the minimum total cost

$$C_i = c_i + \min_{(y_{W_i}) \in A_i} \sum_{j \in W_i} c_j. \tag{3.23}$$

For the conjunctive technological functions, sets (3.21) and (3.22) have the same form $A_n^* = A_n = N$. For the disjunctive technological functions, A_n^* is the set of all nodes of the graph G that lie on the shortest path (in terms of the total cost) from any of its inputs to the output; the value C_n equals the length of this path. In the general case, the methods of graph theory, network scheduling, planning and control can be used.

The Principal considers the interests of different agents, stimulating them to choose required actions in a rather simple way—using the decomposition theorems of the agents' game [24]. Therefore, the only problem that has to be solved by the Principal is to find sets (3.22) (the planning problem).

In addition, it would be interesting to establish general-form sufficient conditions under which the NAS can be *aggregated*, i.e., represented as an equivalent network structure of a single element with constructively defined properties (of course, depending on the properties of the elements of the original network).

The above NAS model is based on the assumption that the Principal completely knows the network G and also all its technological functions. However, technology design often consists in the development of a sequence of actions under a prior uncertainty in the external conditions and also in the knowledge about possible methods for achieving a *goal* (the result of a corresponding agent), i.e., a prior uncertainty in the cause-effect and/or logical relations between different potential technology elements.

The technology design process is (a) to eliminate this uncertainty (considering the cost) using the Principal's purposeful actions that modify his/her beliefs about the NAS structure and (b) to synthesize *an optimal technology*—the NAS for achieving the required goal with the minimum technology design and implementation cost (cost management during the technology's life cycle).

Consider *the optimal learning problem within the technological network*. For a given technological graph, an agent can analyze as many states of the environment (per unit time) as much resources he/she receives from the Principal. This may form the Principal's managerial actions. Assume each state of the environment requires the same time for analysis.

Then the problem can be written in the following general form:

(1) Construct a technological graph.
(2) For each node, partition the set of possible states of the environment into a finite set of non-intersecting subsets and then estimate the probabilities of each subset (in accordance with Sects. 3.1 and 3.2, the uniform partition should be chosen).
(3) Fix the relation between the learning characteristics of the nodes (agents) and the resources.
(4) Find the relation between the characteristics of the entire technological graph and the resources (in particular, see Formulas (3.21)–(3.23)).
(5) Solve the management problem (the resource allocation problem among the network nodes).

Consider a series of models implementing the last step. Denote by $u \leq 1$ the resource quantity, i.e., the share of the states of the environment that are tested per unit time.

For the uniformly distributed probabilities of the states of the environment, the relation between the learning level, time and resource quantity takes the form

$$L(t) = 1 - \exp(-u\,t) \tag{3.24}$$

(see expression 3.1).

In accordance with the aforesaid, for $K \gg 1$ the expected time τ of reaching a required learning level $L_{req} \in [0, 1)$ can be approximated as

$$\tau(u) = -\frac{\ln(1 - L_{req})}{u}. \tag{3.25}$$

Due to the convexity of (3.25) in u, the following result can be easily established.

Proposition 7 *For any relation between the resource quantity and time, there exists a constant value of this quantity under which a required learning level is reached in the same or greater expected time.*

Assume the technology design/adoption process incurs some cost $c(u\,K)$ on the agent. Let the cost function be a strictly monotonically increasing and convex function of the states of the environment tested per unit time (this number of states can be interpreted as the efficiency of allocated computational resources).

In what follows, two special cases will be studied—sequential and parallel technology design.

Sequential technologies design. Consider n technologies with numbers $i \in \overline{1, n}$ that are designed sequentially (in accordance with their numbering) with the same learning level L_{req} for all of them. Then the design time for the entire complex of these technologies is the sum of the design times of partial technologies:

$$T_{max}(u_1, \ldots, u_n) = -\ln(1 - L_{req}) \sum_{i=1}^{n} \frac{1}{u_i}. \tag{3.26}$$

In this case, the minimum design cost is

$$c_{min}(u_1, \ldots, u_n) = c(\max_{i \in \overline{1,n}}\{u_i K_i\}). \tag{3.27}$$

The minimization problem of the design time (3.26) subject to an upper bound C on the cost function (3.27) has the solution

$$u_i = \frac{c^{-1}(C)}{\max_{i \in \overline{1,n}}\{K_i\}}. \tag{3.28}$$

In other words, the same resource quantity is allocated for the optimal design (3.28) of each technology.

From Formulas (3.26) and (3.28) it follows that the complex of sequential technologies can be represented as an aggregate technology with the following relation between the design time and resource quantity:

$$T(C) = -\ln(1 - L_{req})n\frac{\max_{i \in \overline{1,n}}\{K_i\}}{c^{-1}(C)}. \tag{3.29}$$

The inverse problem, which is to find the minimum cost C_{min} of designing the complex of sequential technologies in a given expected time T, has the solution

$$C_{min} = c\left(\frac{-\ln(1 - L_{req})n \max_{i \in \overline{1,n}}\{K_i\}}{T}\right). \tag{3.30}$$

Parallel technologies design. Now, consider n technologies with numbers $i \in \overline{1,n}$ that are designed in parallel with the same learning level L_{req} for all of them. Then the design time for the entire complex of these technologies is the maximum of the design times of partial technologies:

$$T_{min}(u_1, \ldots, u_n) = \frac{-\ln(1 - L_{req})}{\min_{i \in \overline{1,n}}\{u_i\}}. \tag{3.31}$$

In this case, the minimum design cost is

$$c_{max}(u_1, \ldots, u_n) = \sum_{i=1}^{n} c(u_i K_i). \tag{3.32}$$

The minimization problem of the design time (3.31) subject to an upper bound C on the cost function (3.32) has the solution

$$u_i = -T^*_{\min} \ln(1 - L_{\text{req}}),\qquad\qquad(3.33)$$

where $T^*_{\min}$ satisfies the equation

$$\sum_{i=1}^{n} c\left(-\ln(1 - L_{\text{req}})T^*_{\min} K_i\right) = C.\qquad\qquad(3.34)$$

Note that the same resource quantity is allocated for the optimal design (3.33) of each technology, and the design processes of all technologies are completed simultaneously in the time $T^*_{\min}$.

The inverse problem, which is to find the minimum upper bound $C_{\min}$ for the cost of designing the complex of parallel technologies in a given expected time T, has the solution defined by (3.34) with $T^*_{\min} = T$.

From Formulas (3.31) and (3.33) it follows that the complex of parallel technologies can be represented as an aggregate technology. For the linear cost function $c( )$, the analytic relation between the design time and resource quantity of this aggregate technology is given by

$$T(C) = -\ln(1 - L_{\text{req}})\frac{\sum_{i=1}^{n} K_i}{C}.\qquad\qquad(3.35)$$

Thus, a sequential-parallel network diagram of technologies design can be first decomposed into the sequential and parallel elements with the above optimal resource allocations and equivalent aggregate representations (Formulas 3.29 and 3.35). As a result, this network diagram can be written in a simple analytic aggregate form.

Sequential-parallel networks are called aggregable. In accordance with the well-known aggregability criterion, a network is aggregable if it contains bridges. Any network can be transformed into an aggregable one by splitting a series of nodes into new nodes; in this case, the optimal solution of the minimum design time (or cost) problem for the resulting aggregable network will be a lower bound of the corresponding solution for the initial network.

3.4   Transition from Technology Design to Productive Use

Using the established properties of the technology management processes, consider an important problem of technological decision-making: completing the design phase (block 3 in Fig. 1.9) and performing further transition to productive use at the implementation phase of the CA life cycle (cycle b–c in Fig. 1.9).

Assume the actor invests in the creation of his/her CA technology during the design phase for gaining payoffs from its productive use during the implementation phase. At each time during the design phase, the CA effect for the actor is deterministic and negative: he/she bears some cost c_d regardless of a current state of the environment.

At each time t of the implementation phase, one of two possible events occurs as follows.

- The environment takes one of the known states for which the technology has been developed earlier. In this case, the actor gains a payoff v. This event will be denoted by $\xi_t = 1$.
- The environment takes an unknown state ($\xi_t = 0$), and the technology has to be modernized accordingly. In this case, the subject bears the cost c_p without any payoff (here the inequality $c_p > c_d$ is natural because otherwise the design phase makes no economic sense).

During the implementation phase the effect is uncertain and depends on the state of the environment $v\xi_t - c_p(1 - \xi_t) = (v + c_p)\,\xi_t - c_p$. At time t the expected effect depends on the technology's maturity level L_{t-1} reached by this time and can be calculated as

$$V(t) = (v + c_p)\mathrm{E}[\xi_t] - c_p = (v + c_p)\mathrm{Pr}(\xi_t = 1) - c_p = (v + c_p)L_{t-1} - c_p.$$

For such problems, the standard additivity assumption claims that the effect gained on a time interval is the sum of the effects gained at each time of this interval.

Recall that the CA life cycle is implemented as follows (see Chap. 1 and Fig. 1.9). Before each time, the actor chooses between CA design (which incurs the cost c_d) and CA execution. In the latter case, he/she gains the payoff v or bears the cost of modernization c_p, depending on a current state of the environment.

Let d_t be an indicator function for the actor's decisions at each time t with the following values: "0" (CA design) or "1" (CA execution). Then the actor's expected effect on the time interval between times t_1 and t_2 can be written as

$$V(t_1, t_2) = \sum_{\tau=t_1}^{t_2} \left[(vL_{\tau-1} - c_p(1 - L_{\tau-1}))d_\tau - c_d(1 - d_\tau) \right].$$

Then the technological decision problem on the completion of the design phase and transition to the implementation phase of the CA life cycle is to find the decision strategy $\{d_t^*\}$ maximizing the expected effect $V(t_1, t_2)$:

$$\{d_t^*\} = \arg\max_{\{d_t\}} \sum_{\tau=t_1}^{t_2} \left[(v\,L_{\tau-1} - c_p\,(1 - L_{\tau-1}))\,d_\tau - c_d\,(1 - d_\tau) \right].$$

Assume the actor makes sequential decisions at each time $t = \overline{1, T}$ independently of his/her decisions at the past times. In this case, all $\{d_t\}$ are independent of each other.

Then at the current time t the decision strategy has to maximize $V(t, T)$:

$$\max_{\{d_t\}} V(t, T) = \max_{\{d_t\}} \left\{ \sum_{\tau=t}^{T} (v L_{\tau-1} - c_p (1 - L_{\tau-1})) d_\tau - c_d (1 - d_\tau) \right\}. \quad (3.36)$$

Due to the independence of $\{d_t\}$ for different times, the maximum of the sum of the expected effects is the sum of the maximum effects. From expression (3.36) then it follows that

$$\max_{\{d_t\}} V(t, T) = \sum_{\tau=t}^{T} \max_{d_\tau} \left\{ (v L_{\tau-1} - c_p (1 - L_{\tau-1})) d_\tau - c_d (1 - d_\tau) \right\}$$

$$= \max_{d_t} \left\{ (v L_{t-1} - c_p (1 - L_{t-1})) d_t - c_d (1 - d_t) \right\} + \max_{\{d_{t+1}\}} V(t + 1, T).$$

Making trivial transformations and denoting

$$L_{\text{thres}} = \frac{c_p - c_d}{c_p + v}, \quad (3.37)$$

we finally get

$$\max_{\{d_t\}} V(t, T) = -c_d + (v + c_p) \max_{d_t} \left\{ (L_{t-1} - L_{\text{thres}}) d_{t_m} \right\} + \max_{\{d_{t+1}\}} V(t + 1, T).$$

The following result is immediate.

Proposition 8 *The optimal decision strategy that yields the maximum total effect is given by*

$$d_t(L_{t-1}) = \begin{cases} 0 \text{ if } L_{t-1} < L_{\text{thres}}, \\ 1 \text{ if } L_{t-1} \geq L_{\text{thres}}. \end{cases} \quad (3.38)$$

In other words, the optimal strategy (3.38) has a single switch from 0 to 1 (from the design phase to the implementation phase), and the transition condition is determined by the technology's maturity level as follows. While this level is not exceeding the threshold ($L_{t-1} < L_{\text{thres}}$), the actor benefits from technology design through investments; starting from the time t_{reach} of reaching the maturity level L_{thres}, the actor chooses the productive use of the technology (for gaining the payoffs from activity execution), further improving the maturity level in parallel. First, the actor just designs the technology and then redesigns (improves) it in the course of activity execution.

Using Formulas (3.36) and (3.38) in combination with the Wald identity, we will derive an explicit expression of the prior maximum expected effect

$$V^*(0,\ T) = (v + c_p) \sum_{\tau = t_{\text{reach}} + 1}^{T} L_{\tau - 1} - c_d\ t_{\text{reach}} - c_p\ (T - t_{\text{reach}}), \qquad (3.39)$$

where t_{reach} is the expected time of reaching the maturity level L_{thres}.

Substituting the expected maturity level $L_{\tau - 1}$ into (3.39) gives

$$V^*(0,\ T) = (v + c_p) \sum_{\tau = t_{\text{reach}}}^{T-1} \left(1 - \sum_{k=1}^{K} p_k (1 - p_k)^{\tau} \right) - c_d\ t_{\text{reach}} - c_p\ (T - t_{\text{reach}}).$$

In the final analysis,

$$V^*(0,\ T) = vT - (v + c_d)\ t_{\text{reach}} - (v + c_p)$$
$$- (v + c_p) \sum_{k=1}^{K} \left[(1 - p_k)^{t_{\text{reach}}} - (1 - p_k)^{T} \right].$$

Interestingly, the optimal strategy is independent of the interval length T, which actually defines the resulting effect $V^*(0,\ T)$ only.

The following conclusion is immediate from (3.37) and the monotonic increase of the process L_t (see Proposition 2 and its extensions to different integration processes). For any arbitrarily large cost c_d and c_p such that $c_d < c_p$ and any arbitrarily small nonzero payoff v, there exists a time T_{payb} since which the agent's activity will yield a positive effect. In other words, T_{payb} will determine the break-even point of the CA life cycle. This time can be found from the equation

$$a_1\ t + \sum_{k=1}^{K} (1 - p_k)^t = a_2\ t_{\text{reach}} + 1 + \sum_{k=1}^{K} (1 - p_k)^{t_{\text{reach}}}$$

in t, where $0 < a_1 = v\ /(v + c_p) < 1$ and $0 < a_2 = (v + c_d)/(v + c_p) < 1$. The monotonic increase of the technology's maturity level (see Proposition 2 and its extensions to different integration processes) allows proving that the sequential single-switch strategy is optimal among all sequential decision strategies d_t (not only among the strategies with independent decisions at each time). On the other hand, the optimal sequential decision strategy is not worse than any prior strategy. This leads to another important result as follows.

Proposition 9 *The sequential single-switch strategy (3.38) is optimal among all admissible ones. The resulting effect (3.39) is maximum achievable while the payback time T_{payb} is minimum possible.*

Now, consider the technological decision problem on the transition from the design phase to the implementation phase under the unknown but fixed characteristics of the environment (the dimension K and also the probabilities $\{p_k\}$). In this case, the technology's maturity level cannot be calculated and hence the decision strategy (3.38) (see Proposition 6) becomes inapplicable.

The expected effect can be written as

$$\max_{\{d_t\}} V(t, T) = -c_d + (v + c_p) \max_{d_t}\{(\Pr(\xi_t = 1) - L_{\Pi op})d_t\} + \max_{\{d_{t+1}\}} V(t + 1, T).$$

Therefore, the sequential strategy optimizing the expected effect $V(t, T)$ consists in $(\Pr(\xi_t = 1) - L_{thres})d_t \to \max_{d_t}$, which is achieved by $d_t = 1$ if $\Pr(\xi_\tau = 1) > L_{thres}$ and $d_t = 0$ otherwise. In other words, d_t must be the result of the sequential testing of the composite main hypothesis $L_t < L_{thres}$ against its composite alternative $L_t \geq L_{thres}$ (whether the value of the unobserved process L_t is exceeding the threshold L_{thres} or not).

Under the unknown characteristics of the environment, all available information for decision-making is whether a current state has been observed before or not. Denote by θ_k the times when the environment takes new states never observed before. These times form an increasing finite sequence $0 = \theta_1 < \theta_2 < \ldots < \theta_k < \ldots < \theta_K$ observed by the actor. By definition, at each time θ_k the process L_t has an unknown increment p_k and takes the value $L_{\theta_k} = \sum_{i=1}^{k} p_i$, which is fixed till the next time θ_{k+1}. Consider the series lengths $\psi_k = (\theta_{k+1} - \theta_k - 1)$ for $k = \overline{1, K - 1}$. The values ψ_k are independent random variables, each obeying the geometric distribution parameterized by the current partial sum $L_{\theta_k} = \sum_{i=1}^{k} p_i$ of the probabilities of all realized states of the environment. (Note that this parameter is unknown to the actor.) In other words, $\Pr(\psi_k = n) = (1 - L_{\theta_k})L_{\theta_k}^{n-1}$, and the expected values and variances are $L_{\theta_k}/(1 - L_{\theta_k})$ and $L_{\theta_k}/(1 - L_{\theta_k})^2$, respectively. The actor has no prior knowledge about the dimension K and the distribution $\{p_k\}$; hence, the length of the sequence $\{\psi_k\}$ cannot be defined and this sequence should be considered a priori infinite.

Let s be the number of the last new state observed by the current time t, i.e., $0 = \theta_1 < \theta_2 < \ldots < \theta_s \leq t$. Also introduce the notation $\psi_s = t - \theta_s$.

Thus, at each time t the actor disposes of the following information for his/her decisions:

- the series lengths $\psi_1, \psi_2, \ldots, \psi_{s-1}, \psi_s$, further denoted by $\{\psi\}$;
- the knowledge that each of ψ_k is generated by the geometric distribution with an unknown increasing parameter $L_{\theta_k} < L_{\theta_{k+1}}$.

Under the unknown properties of the environment, the problem is to synthesize a sequential testing criterion $d_t(\{\psi\})$ for the composite main hypothesis H_0: $L_{\theta_s} < L_{thres}$ (the value L_t has not exceeded the threshold L_{thres} before the time t) against the set of composite alternatives $\{H_i : L_{\theta_i} \geq L_{thres}\}$ (the value L_t has exceeded the threshold L_{thres} at the time θ_i) by maximizing the expected effect $V(0, T)$. For the synthesis procedure of this criterion, the decision strategy $d_t(\{\psi\})$ will be chosen using the likelihood ratio while the criterion parameters will be adjusted by optimizing the expected effect $V(0, T)$.

The relative likelihood ratio of the series lengths $\{\psi\}$ generated by the geometric distribution has the form

$$l(i, t) = \ln\left\{\frac{\Pr(\{\psi\}|H_i)}{\Pr(\{\psi\}|H_0)}\right\} = \sum_{k=i}^{s-1} \ln\left(\frac{L_k^i}{L_k^0}\right)\psi_k + \sum_{k=i}^{s-1} \ln\left(\frac{1 - L_k^i}{1 - L_k^0}\right) + \ln\left(\frac{L_s^i}{L_s^0}\right)\psi_s.$$

where L_k^0 and L_k^i give the values of the process L_t at the times θ_k when the main $L_k^0 < L_{\text{thres}}$ and alternative $L_k^i \geq L_{\text{thres}}$ hypotheses are true, respectively, and i ($i = 1, 2, \ldots s$) means the number of the alternative hypothesis (actually, the serial number of the new ith state of the environment).

There is no available information on the properties of the environment, and the only constructive considerations about the values L_k^0 and L_k^i are the inequalities $L_k^0 < L_{\text{thres}}$ and $L_k^i \geq L_{\text{thres}}$. Hence, let $L_k^0 = L_{\text{thres}} - \Delta L$ and $L_k^i = L_{\text{thres}} + \Delta L$. (As a matter of fact, this assumption seems neither better nor worse than any other.)

At a current time t, the main hypothesis will be rejected for its alternative if at least one of the functions $l(i, t)$ exceeds some threshold l_{thres}, i.e., $\max_i l(i, t) \geq l_{\text{thres}}$.

Denote $l(t) = \max_i l(i, t)$ and study how this function will vary with the course of time. If at a current time t a known state of the environment is observed, then each of the likelihoods $l(i, t)$, $1 \leq i \leq s$, will increase by $a_1 = \ln\left(\frac{L_{\text{thres}} + \Delta L}{L_{\text{thres}} - \Delta L}\right) > 0$. As a result, $l(t)$ will increase by the same value, $l(t + 1) = l(t) + a_1$. If a new state is observed, then each of $l(i, t)$, $1 \leq i \leq s$, will decrease by $a_2 = \ln\left(\frac{1 - L_{\text{thres}} - \Delta L}{1 - L_{\text{thres}} + \Delta L}\right) < 0$. Also the new $(s + 1)$th function $l(s + 1, t + 1) = 0$ will be formed. Therefore, in this case $l(t + 1) = \max\{0; l(t) + a_2\}$.

The resulting value of the likelihood function $l(t)$ is compared with the threshold l_{thres}. Because all the three constants a_1, a_2 and l_{thres} have to be determined, without loss of generality let $a_1 = 1$ and $a_2 = -a$.

Thus, we have derived the sequential likelihood-based criterion to be used under the unknown properties of the environment. This criterion includes the iterative calculation of the likelihood function $l(t)$ and the corresponding decision strategy $d_t(\{\psi\})$ given by

$$l(t + 1) = \begin{cases} l(0) = 0, \\ l(t) + 1 \text{ if known state is observed}, \\ \max\{0; l(t) - a\} \text{ if new state is observed}; \end{cases}$$

$$d_t(\{\psi\}) = \begin{cases} 0 \text{ if } d_{t-1}(\{\psi\}) = 0 \text{ and } l(t) < l_{\text{thres}}, \\ 1 \text{ if } d_{t-1}(\{\psi\}) = 1 \text{ or } l(t) \geq l_{\text{thres}}. \end{cases} \quad (3.40)$$

For completing the synthesis procedure, it remains to calculate the constants a and l_{thres} by optimizing the expected effect $V(0, T, a, l_{\text{thres}})$ under different assumptions of the environment's properties.

To this end, write the expected effect on some time interval between 0 and T (with the initial learning level $L_0 = 0$) as a function of the constants a and l_{thres} and also of the assumed properties of the environment—the distribution $\{p_k\}$ and its dimension K.

Hereinafter, a sequence of the numbers $\{k_1, k_2, \ldots, k_K\}$ of new states of the environment in the order of their observation during technology design will be called a trajectory while the times when the states k_i occur will be denoted by θ_i. As a trajectory evolves, the learning level varies from 0 to 1, taking the values $L_{\theta_i} = \sum_{j=1}^{i} p_{k_i}$ at the times θ_i. The probability of each trajectory $\{k_i\}$ is

$$P(\{k_i\}) = \prod_{i=1}^{K} \left[p_{k_i} \left(1 - L_{\theta_i - 1}\right)^{-1} \right].$$

First of all, calculate the effect $V(\{k_i\}, T, a, l_{\text{thres}})$ for each trajectory and then perform averaging over all trajectories:

$$V(T, a, l_{\text{thres}}) = \sum_{\{k_i\}} V(\{k_i\}, T, a, l_{\text{thres}}) P(\{k_i\})$$

$$= \sum_{\{k_i\}} V(\{k_i\}, T, a, l_{\text{thres}}) \prod_{i=1}^{K} \left[p_{k_i} \left(1 - L_{\theta_i - 1}\right)^{-1} \right]. \qquad (3.41)$$

For calculating the effect $V(\{k_i\}, T, a, l_{\text{thres}})$, each trajectory $\{k_i\}$ is assumed to be fixed. Hence, for the sake of simplicity let $k_i = i$, and accordingly $p_i = p_{k_i}$ and $L_i = L_{\theta_i}$.

Introduce a two-dimensional discrete random process $(k(t); l(t))$, where $k(t)$ denotes the number of all unrealized trajectory states at time t ($k(0) = K - 1$, and at the subsequent times θ_i the function $k(t)$ is gradually decreasing by 1 down to 0); $l(t)$ denotes the values of the likelihood function ($l(0) = 0$, and at the subsequent times $l(t)$ is varying between 0 and l_{thres} inclusive in accordance with rule (3.40)).

If at some time t the second component of the process $(k(t), l(t))$ reaches the value $l(t) = l_{\text{thres}}$, then the effect for this trajectory will take the value $v(T - t) - c_p k(t) - c_d t$.

Therefore, the expected effect for the trajectory is

$$V(\{k_i\}, T, a, l_{\text{thres}})$$

$$= \sum_{t; k(t)} \left(v(T - t) - c_d t - c_p k(t) \right) \Pr(k(t); l_{\text{thres}}; t)$$

$$= v\, T - (c_d + v) \left\{ \sum_{t; k(t)} t \Pr(k(t); l_{\text{thres}}; t) + \frac{c_p}{c_d + v} \sum_{t; k(t)} k(t) \Pr(k(t); l_{\text{thres}}; t) \right\}.$$

Consider this expression in detail. The first term vT depends neither on the properties of the environment, nor on the trajectory and nor on the parameters of the criterion. Hence, it will be omitted for the sake of simplicity. The second term describes the learning cost: the constant $(v + c_d)$ is multiplied by the sum of the expected time of reaching the required learning level (the first sum) and the expected number of unrealized states of the environment (the second sum) with the factor $\mu = c_p (v + c_d)^{-1}$.

Thus, the optimal effect can be calculated by minimizing the cost

$$C(\{k_i\}, T, a, l_{\text{thres}}) = \sum_{t;\, k(t)} t \Pr(k(t);\, l_{\text{thres}};\, t) + \mu \sum_{t;\, k(t)} k(t) \Pr(k(t);\, l_{\text{thres}};\, t).$$

(3.42)

In view of (3.42), effect (3.41) takes the form

$$V(T, a, l_{\text{thres}}) = v\, T - (c_d + v) \sum_{\{k_i\}} C(\{k_i\}, T, a, l_{\text{thres}}) P(\{k_i\})$$

$$= v\, T - (c_d + v) \sum_{\{k_i\}} C(\{k_i\}, T, a, l_{\text{thres}}) \prod_{i=1}^{K} \left[p_{k_i} \left(1 - L_{\theta_i - 1}\right)^{-1} \right].$$

This means that the effect optimization problem $V(0, T, a, l_{\text{thres}}) \to \max$ is equivalent to the expected cost minimization problem $C(T, a, l_{\text{thres}}) \to \min$, where

$$C(T, a, l_{\text{thres}}) = \sum_{\{k_i\}} C(\{k_i\}, T, a, l_{\text{thres}}) \prod_{i=1}^{K} \left[p_{k_i} \left(1 - L_{\theta_i - 1}\right)^{-1} \right]$$

$$= \sum_{\{k_i\}} \left(\sum_{t;\, k(t)} (t + \mu\, k(t))\, t\; \Pr(k(t);\, l_{\text{thres}};\, t) \right) \prod_{i=1}^{K} \left[p_{k_i} \left(1 - L_{\theta_i - 1}\right)^{-1} \right]. \quad (3.43)$$

For solving this problem, we have to find the probability distribution $Pr(k(t);\, l_{\text{thres}};\, t)$.

Also note that the expected cost can be written as

$$C(T;\, a;\, l_{\text{thres}})$$

$$= \sum_{\{k_i\}} C(\{k_i\};\, T;\, a;\, l_{\text{thres}}) P(\{k_i\})$$

$$= \sum_{\{k_i\}} \left(\sum_{t;\, k(t)} (t + \mu\, k(t))\; \Pr(k(t);\, l_{\text{thres}};\, t) \right) P(\{k_i\})$$

$$= \sum_{\{k_i\}\, t;\, k(t)} t\; \Pr(k(t);\, l_{\text{thres}};\, t) P(\{k_i\}) + \mu \sum_{\{k_i\}\, t;\, k(t)} k(t)\; \Pr(k(t);\, l_{\text{thres}};\, t) P(\{k_i\}).$$

Consequently,

$$C(T;\, a;\, l_{\text{thres}}) = \bar{t}(K;\, \{p_k\};\, a;\, l_{\text{thres}}) + \mu \bar{k}(K;\, \{p_k\};\, a;\, l_{\text{thres}}), \quad (3.44)$$

where $\bar{t}(K;\, \{p_k\};\, a;\, l_{\text{thres}})$ gives the expected time of reaching the level l_{thres} by the process $l(t)$; $\bar{k}(K;\, \{p_k\};\, a;\, l_{\text{thres}})$ is the expected number of unrealized states of the environment by the time when the process $l(t)$ reaches the level l_{thres}; finally, the known parameter $\mu = c_p(v + c_d)^{-1}$ characterizes the general impact of the payoff v and cost c_p and c_d.

Formula (3.44) can be used for the qualitative analysis of the expected cost dynamics under different values of the parameters. Obviously, the function $\bar{t}(\cdot)$ is monotonically increasing in a and l_{thres} while the function $\bar{k}(\cdot)$ is monotonically decreasing in a and l_{thres} for any probability distributions of all states of the environment. For $l_{\text{thres}} = 0$ (obviously, $a \leq l_{\text{thres}}$), the criterion will be satisfied immediately at time 1; the functions $\bar{t}(\cdot)$ and $\bar{k}(\cdot)$ will be equal to 1 and $K - 1$, respectively, and the expected cost will become $C(T, 0, 0) = 1 + \mu (K - 1)$. Conversely, for "very large" values l_{thres}, the criterion will not be satisfied on the entire interval to T, the functions $\bar{t}(\cdot)$ and $\bar{k}(\cdot)$ will be equal to T and 0, respectively, and the expected cost will become $C(T, a, \infty) = T$ (for any a).

Hence, the expected cost $C(T, a, l_{\text{thres}})$ has an optimum that depends on K, $\{p_i\}$, μ, a, and l_{thres} (in a special case, the optimum corresponds to one of the limit values, $1 + \mu (K - 1)$ or T).

By definition the process $(k(t), l(t))$ evolves in accordance with the following rules.

- At the initial time $t = 1$, the process has the deterministic value $(k(0); l(0)) = (K - 1; 0)$.
- At each time $t > 0$ when a known state of the environment is realized again, $(k(t); l(t)) = (k(t - 1); l(t - 1) + 1)$. If also $l(t - 1) \geq l_{\text{thres}}$, then $(k(t; l(t)) = (k(t - 1); l_{\text{thres}})$. The probability of this event is $L_{k(t)}$.
- At each time $t > 0$ when an unknown state of the environment is realized, $(k(t); l(t)) = (k(t - 1) + 1; \max\{0; l(t - 1) - a\})$. If also $l(t - 1) \geq l_{\text{thres}}$, then $(k(t); l(t)) = (k(t - 1) + 1; l_{\text{thres}})$. The probability of this event is $1 - L_{k(t)}$.

For $0 \leq k < K$ and $0 < l < l_{\text{thres}}$, the evolution of the probability function $P(k; l; t)$ of the process $(k(t), l(t))$ with the discrete time $t > 0$ can be therefore described by the system of difference equations

$$P(k; l; t) = L_k P(k; l-1; t-1) + (1 - L_{k+1}) P(k + 1; l + a; t-1), \qquad (3.45)$$

with the initial conditions

$$P(K-1; 0; 0) = 1 \text{ and } P(k; l; 0) = 0 \text{ for any } k < K-1 \text{ or } l > 0, \qquad (3.46)$$

and with the boundary-value conditions

$$P(k; l_{\text{thres}}; t) = L_k(P(k; l_{\text{thres}} - 1; t - 1) + P(k; l_{\text{thres}}; t - 1)) \\ + (1 - L_{k+1}) P(k + 1; l_{\text{thres}}; t - 1),$$

$$P(k; 0; t) = \sum_{l=1}^{a} (1 - L_{k+1}) P(k + 1; l; t - 1),$$

$$P(k; l; t) \equiv 0 \quad \text{if } l_{\text{thres}} < l \text{ or } l < 0 \; \forall k, \forall t. \qquad (3.47)$$

Using the system of difference Eqs. (3.45)–(3.47), we may iteratively calculate the probability function of the process $(k(t); l(t))$. In turn, the probability distribution of the times of reaching the threshold level l_{thres} and the number of the unrealized trajectory states can be obtained as $Pr(k(t); l_{\text{thres}}; t) = L_k P(k; l_{\text{thres}} - 1; t - 1)$. This gives the expected cost, and hence the cost $C(T, a, l_{\text{thres}})$ can be optimized numerically by choosing the optimal values of the parameters a and l_{thres} so that $C(T, a, l_{\text{thres}})$ → min for the expected properties of the environment—the probability distribution $\{p_k\}$ and its dimension K.

Thus, we have derived the sequential likelihood-based criterion (3.40) to be used under the unknown properties of the environment (the dimension K and the probability distribution $\{p_k\}$). For this criterion, an algorithm of choosing the optimal parameter values (in terms of the minimum cost) under the expected properties of the environment has been suggested.

The properties of the synthesized criterion were examined by simulation modeling for the numerical solution of the difference Eqs. (3.45)–(3.47) and the iterative calculation of the probability function $Pr(k(t); l_{\text{thres}}; t)$ and the expected cost $C(T, a, l_{\text{thres}})$. The optimal values of the parameters a and l_{thres} were chosen by the exhaustive search with a given step depending on the expected properties of the environment—the probability distribution $\{p_k\}$ and its dimension K.

Different scenarios were considered as follows: the uniformly distributed probabilities of the states of the environment of the dimension $K = 5, \ldots, 80$ varied with step 5 and the parameter $\mu = 3, \ldots, 30$ varied with step 3. The simulation results led to the following conclusions.

First, for all the values K and μ considered, the least expected cost was achieved for $a = 0$; see Tables 3.1, 3.2, 3.3 and 3.4. This result will be interpreted below.

Second, the following functions were constructed in tabular form (see Tables 3.5 and 3.6):

- the minimum in l_{thres} expected cost $C_{opt}(K; \mu)$ for different K and μ, $a = 0$;
- the optimal value $l_{opt}(K; \mu)$ of parameter l_{thres} for different K and μ under which the minimum cost $C_{opt}(K; \mu)$ is achieved (see Table 3.4).

Table 3.1 Expected cost for $K = 8$, $\mu = 3$, $l_{\text{thres}} = 6, \ldots, 11$ and $a = 0, \ldots, (l_{\text{thres}} - 1)$

	6	...	9	10	11	...	19	20
0	14.2	...	13.62	13.61	13.66	...	15.5	16.1
1	13.9	...	14.2	14.4	14.6	...	17.6	18.0
2	14.1	...	14.8	15.1	15.4	...	18.7	19.2
...			...	...	...	...	...	...
10					17.1	...	21.8	22.3
11						...	21.9	22.4
12						...	22.0	22.5
...						...	...	...
19								22.8

Table 3.2 Expected cost for $K = 8$, $\mu = 27$, $l_{thres} = 20, \ldots, 35$ and $a = 0, \ldots, (l_{thres} - 1)$

	20	...	28	29	30	...	35
0	27.79	...	24.36	23.98	24.33	...	24.63
1	25.4	...	24.7	24.9	25.1	...	26.4
2	25.0	...	25.4	25.6	25.9	...	27.5
...	...	...	...	...	...	...	...
19	26.3	...	28.7	29.1	29.5	...	31.7
20			28.8	29.2	29.6	...	31.8
...			...	...	...	...	...
27			28.9	29.3	29.7	...	32.0
28				29.3	29.7	...	32.0
...						...	...
34							32.0

Table 3.3 Expected cost for $K = 60$, $\mu = 3$, $l_{thres} = 55, \ldots, 61$ and $a = 0, \ldots, (l_{nop} - 1)$

	55	56	57	58	59	60	61
0	83.45	83.42	83.41	83.41	83.42	83.44	83.47
1	87.7	88.0	88.2	88.5	88.8	89.1	89.4
2	93.4	93.8	94.2	94.6	95.0	95.3	95.7
3	98.3	98.8	99.2	99.6	100.1	100.5	101.0
...	...	...	...	...	...	...	...
54	134.7	135.7	136.6	137.6	138.5	139.5	140.4
...						...	...
59						139.5	140.5
60							140.5

Table 3.4 Expected cost for $K = 60$, $\mu = 27$, $l_{thres} = 170, \ldots, 188$ and $a = 0, \ldots, (l_{thres} - 1)$

	170	...	180	181	182	...	187	188
0	149.35	...	148.91	148.87	148.92	...	149.07	149.20
1	151.96	...	154.10	154.35	153.64	...	155.88	156.15
2	157.32	...	160.37	160.69	159.73	...	162.69	163.03
3	157.98	...	165.65	166.01	164.92	...	168.24	168.63
...	...	...	...	...	...	...	...	...
180				225.8	226.4	...		
181					226.4	...		
...						...	...	...
186							229.1	229.6
187								229.6

Table 3.5 Minimum in l_{thres} expected cost $C_{opt}(K; \mu)$ for different K and μ (minimum is achieved at $l_{thres} = l_{opt}(K; \mu)$, see Table 3.6)

	3	6	9	12	15	18	21	24	27	30
5	7	8	9	10	10	11	11	11	11	12
10	14	17	19	20	21	22	23	23	24	24
15	21	26	29	31	32	34	35	36	36	37
20	28	34	38	41	43	45	47	48	49	50
25	35	43	48	52	54	57	58	60	61	63
30	42	52	58	62	65	68	70	72	74	76
35	49	61	68	72	76	79	82	84	86	88
40	55	69	77	83	87	91	94	97	99	101
45	62	78	87	93	98	102	106	109	111	114
50	69	87	97	104	109	114	118	121	124	127
55	76	95	106	114	120	125	130	133	136	139
60	83	104	116	125	131	137	141	145	149	152
65	90	113	126	135	142	148	153	158	161	165
70	97	122	136	146	153	160	165	170	174	177
75	104	130	145	156	164	171	177	182	186	190
80	111	139	155	167	175	183	189	194	199	203

Table 3.6 Optimal value $l_{opt}(K; \mu)$ of parameter l_{thres} for different K and μ under which minimum cost $C_{opt}(K; \mu)$ is achieved (see Table 3.4)

	3	6	9	12	15	18	21	24	27	30
5	6	8	10	10	12	12	14	14	14	14
10	10	15	19	23	25	25	27	29	29	31
15	14	24	30	34	36	38	42	42	44	46
20	19	31	39	45	49	51	55	57	59	61
25	24	39	48	56	60	64	68	72	74	78
30	29	47	59	67	73	79	83	87	89	93
35	34	55	68	78	84	92	96	100	104	108
40	38	63	77	89	97	105	111	115	119	123
45	43	70	87	100	110	118	124	130	134	140
50	48	78	97	111	121	131	137	145	149	155
55	53	86	107	122	134	144	152	158	164	170
60	58	94	117	133	145	157	165	173	179	187
65	62	102	126	144	158	170	180	188	194	202
70	67	110	136	155	170	183	193	201	209	217
75	72	118	146	166	182	196	206	216	226	232
80	76	125	155	177	195	209	221	231	241	249

In accordance with Tables 3.5 and 3.6 and also the graphs in Figs. 3.7 and 3.8, for any fixed μ both functions $C_{opt}(K; \mu)$ and $l_{opt}(K; \mu)$ can be considered linear in K with a high accuracy for engineering applications ($R^2 \geq 0.99$). Their linear approximations have the form

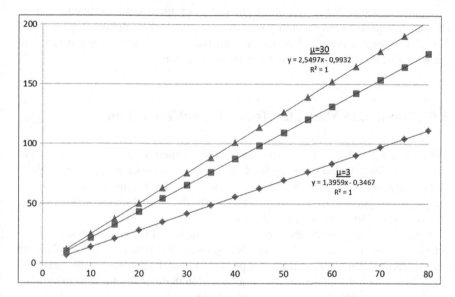

Fig. 3.7 Graph of $C_{opt}(K; \mu)$ for different K and fixed μ

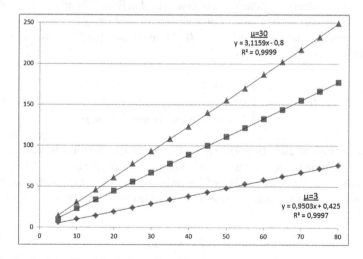

Fig. 3.8 Graph of $l_{opt}(K; \mu)$ for different K and fixed μ

$$C_{opt}(K; \mu) = K \cdot (0.5 \cdot \ln(\mu) + 0.85) - 0.33 \cdot \ln(\mu) + 0.13,$$
$$l_{opt}(K; \mu) = K \cdot (0.95 \cdot \ln(\mu) - 0.12) - 0.05 \cdot \mu + 0.56. \tag{3.48}$$

Expressions (3.48) can be used to perform preliminary calculations and also to choose the initial values of the parameter l_{thres} in simulations.

Approximations (3.48) well match the results of numerical experiments and, at the same time, have a high accuracy (see the values R^2 for the graphs in Figs. 3.7 and 3.8). Such a high accuracy of the linear approximations of experimental data needs theoretical study, which will be done in the next section.

3.5 Simulation Model for Technological Transition

Tables 3.1, 3.2, 3.3 and 3.4 present the values of criterion (3.40) obtained by the simulations for calculating the optimal values of the parameters l_{thres} and a. Each of the tables contains the values of the expected cost, the parameters l_{thres} and a are varying in columns and rows, respectively. Clearly, the optimal values of the expected cost correspond to the upper row, i.e., $a = 0$.

Next, Table 3.5 presents the expected cost calculated by the simulations for different dimensions of the probability distribution of the states of the environment and different parameters μ. Table 3.6 presents the optimal value of the threshold l_{thres} under which the minimum cost is achieved. In these tables, the dimension K is varying in rows while the parameter μ in columns.

The graphs of the expected cost $C_{opt}(K; \mu)$ and the optimal threshold $l_{opt}(K; \mu)$ as functions of K under fixed value μ are demonstrated in Figs. 3.7 and 3.8, respectively.

These results should be thoroughly analyzed for explaining trends (3.48).

For the optimal parameter value $a = 0$, the likelihood functions $l(t)$ of criterion (3.40) are constructed as follows:

$$l(t + 1) = \begin{cases} l(0) = 0, \\ l(t) & \text{if new state is observed,} \\ l(t) + 1 & \text{if known state is observed.} \end{cases} \tag{3.49}$$

Consequently, at each time (in particular, when the criterion is satisfied—the likelihood function $l(t)$ reaches the threshold l_{thres}) we have the equality $l(t) = t - (K - k(t))$, where t is the total number of observations and $(K - k(t))$ is the number of observations with newly realized states of the environment never occurring before. As a result,

$$t = l(t) + K - k(t). \tag{3.50}$$

The expected value of $(K - k(t))$ is the learning level multiplied by K, i.e., $K(1 - (1 - 1/K)^t)$. On the other hand, the expected value of $k(t)$ (the number of all unrealized states) is $E[k(t)] = K(1 - 1/K)^t$. Hence, it follows that the expected time $\bar{t}(K; \{p_k\}; a; l_{\text{thres}})$ of reaching the level l_{thres} by the process $l(t)$ and the expected number $\bar{k}(K; \{p_k\}; a; l_{\text{thres}})$ of the unrealized states of the environment by this time satisfy the equation

$$\bar{k}(K; l_{\text{thres}}) = K(1 - 1/K)^{\bar{t}(K;l_{\text{thres}})}. \tag{3.51}$$

Hereinafter, $\bar{t}(K; l_{\text{thres}}) = \bar{t}(K; \{p_k\}; a; l_{\text{thres}})$ for the case $a = 0$.
The expected cost (3.44) are minimized by choosing a value l_{thres} such that

$$\frac{\partial}{\partial l_{\text{thres}}} C(T; a; l_{\text{thres}}) = \frac{\partial}{\partial l_{\text{thres}}} \bar{t}(K; l_{\text{thres}}) + \mu \frac{\partial}{\partial l_{\text{thres}}} \bar{k}(K; l_{\text{thres}}) = 0 \tag{3.52}$$

Substituting (3.51) into (3.52) gives the equation

$$\left(1 + \mu \ln(1-1/K) K (1-1/K)^{\bar{t}(K;l_{\text{opt}})}\right) \frac{\partial}{\partial l} \bar{t}(K; l_{\text{opt}}) = 0. \tag{3.53}$$

From this equation, find the expected time $\bar{t}(K; l_{\text{opt}})$ corresponding to the optimal expected cost and then, using $\bar{t}(K; l_{\text{opt}})$, calculate the optimal threshold l_{opt}.

The derivative $\frac{\partial}{\partial l} \bar{t}(K; l)$ never vanishes because the expected time is monotonically increasing in l. Therefore, the desired value $\bar{t}(K; l_{\text{opt}})$ can be obtained from

$$1 + \mu \ln(1-1/K) K (1-1/K)^{\bar{t}(K;l_{\text{opt}})} = 0.$$

Solve this equation in $\bar{t}(\cdot)$, taking into account $\ln(1-1/K) \approx -1/K$. The value $\bar{t}(K; l_{\text{opt}})$ corresponding to the optimal expected cost is

$$\bar{t}(K; l_{\text{opt}}) \approx K \ln(\mu). \tag{3.54}$$

Using (3.54) and (3.55), find the approximate relation of $\bar{t}(K; l)$, $\bar{k}(K; l)$ and l_{opt}. Apply the expectation operator to (3.50):

$$E[t] = E[l(t)] + K - E[k(t)].$$

In view of (3.51), write

$$E[t] = E[l(t)] + K - K(1 - 1/K)^{E[t]}.$$

Let $E[l(\bar{t}(K; l))] = l$ to obtain

$$\bar{t}(K; l) \approx l + K - K(1 - 1/K)^{\bar{t}(K;l)},$$

or, in an equivalent form,

$$l \approx \bar{t}(K; l) - K + K(1 - 1/K)^{\bar{t}(K; l)}. \tag{3.55}$$

Finally, substitute (3.54) into (3.55) to get

$$
\begin{aligned}
l_{\text{opt}} &\approx \bar{t}(K; l) - K + K(1 - 1/K)^{\bar{t}(K; l)} \\
&= K \ln(\mu) - K + K(1 - 1/K)^{K \ln(\mu)} \\
&= K \ln(\mu) - K + K e^{K \ln(1 - 1/K) \ln(\mu)} \\
&\approx K \ln(\mu) - K + K e^{-\ln(\mu)} = K \ln(\mu) - K + K/\mu. \tag{3.56}
\end{aligned}
$$

The approximate value of the optimal threshold (3.56) is $l_{\text{opt}} \approx K \ln(\mu) - K + K/\mu$.

Due to the linear relation of $\bar{t}(K; l)$, $\bar{k}(K; l)$ and l_{thres} (see expression 3.50), the optimal expected cost will have a similar trend:

$$C(T; \ a = 0; \ l_{\text{opt}}) \sim K \ln(\mu) + K. \tag{3.57}$$

Thus, the analytic approximations of the optimal threshold (3.56) and expected cost (3.57) well reflect the main trends of the simulation experiments (3.48).

In this chapter, a set of management problems for the design and adoption of complex activity technologies has been considered. Using the optimal learning problem, it has been demonstrated that the uniform partition of the set of all possible states of the environment is asymptotically optimal in terms of the minimum expected error and entropy and also in terms of the maximum expected learning level. For the resource allocation problem in aggregable technological networks, the optimal resource allocation procedures have been obtained in simple analytic form. For the technological decision problem on the transition from design to productive use, the optimal single-switch time has been estimated.

Promising lines of further research include the analytic methods for solving the optimal resource allocation problems for the large classes of technological networks and also the optimal learning problems for more complicated models (in particular, with nonstationary probability distributions of all possible states of the environment; their dependence on accumulated experience and the interaction of different agents; the dependence of cost functions on the realized states of the environment, etc.).

References

1. Novikov D (2012) Collective learning-by-doing. IFAC Proc Vol 45(11):408–412
2. Ebbinghaus H (1885) Über das Gedächtnis. Dunker, Leipzig, 168 pp
3. Hull C (1943) Principles of behavior and introduction to behavior theory. D. Appleton Century Company, New York, 422 pp
4. Stenberg S (1963) Stochastic learning theory. In: Handbook on mathematical psychology, vol I. Wiley, New York, pp 1–120
5. Thurstone L (1919) The learning curve equation. Psychol Monogr 26(3):1–51

6. Thurstone L (1930) The learning function. J Gen Psychol 3:469–493
7. Tolman E (1934) Theories of learning. In: Moss FA (ed) Comparative psychology. Prentice Hall, New York, pp 232–254
8. Anzanello M, Fogliatto F (2011) Learning curve models and applications: literature review and research directions. Int J Ind Ergon 41:573–583
9. Donner Y, Hardy J (2015) Piecewise power laws in individual learning curves. Psychon Bull Rev 22:1308–1319
10. Jaber M (2017) Learning curves: theory, models and applications. CRC Press, Boca Raton, 476 pp
11. Novikov D (1998) Laws of iterative learning. Trapeznikov Institute of Control Sciences RAS, Moscow, 98 pp (in Russian)
12. Aumann R (2008) Rule-rationality versus act-rationality. Discussion paper no. 497, Hebrew University, Jerusalem, 20 pp
13. Foerster H (1995) The cybernetics of cybernetics, 2nd edn. Future Systems, Minneapolis, 228 pp
14. Belov M, Novikov D, Methodology of complex activity. Lenand, Moscow, 320 pp (in Russian)
15. Goertzel B, Iklé M, Goertzel I, Heljakka A (2008) Probabilistic logic network. Springer, Heidelberg, 333 pp
16. Richardson M, Domingos P (2006) Markov logic networks. Mach Learn 62:107–136
17. Kohut R, Steinbach B (2014) Decomposition of boolean function sets for boolean neural networks. https://www.researchgate.net/publication/228865096_Decomposition_of_Boolean_Function_Sets_for_Boolean_Neural_Networks
18. Brachman R, Levesque H (2004) Knowledge representation and reasoning. Morgan Kaufmann, New York, 381 pp
19. Handbook of knowledge representation. Elsevier, Amsterdam, 1034 pp
20. Lucio-Arias D, Scharnhorst A (2012) Mathematical approaches to modeling science from an algorithmic-historiography perspective. In: Scharnhorst A, Börner K, van den Besselaar P (eds) Models of science dynamics. Understanding complex systems. Springer, Heidelberg, pp 23–66
21. Vitanov N, Ausloos M (2012) Knowledge epidemics and population dynamics models for describing idea diffusion. In: Scharnhorst A, Börner K, van den Besselaar P (eds) Models of science dynamics. Understanding complex systems. Springer, Heidelberg, pp 69–125
22. Sauser B, Magnaye R, Tan W, Ramirez-Marquez J (2010) Optimization of system maturity and equivalent system mass for space systems engineering management. In: Proceedings of the Conference on Systems Engineering Research, Hoboken, NJ, March 2010, p 10
23. Sauser B, Ramirez-Marquez J (2011) Development of systems engineering maturity models and management tools. Report no. SERC-2011-TR-014, Stevens Institute of Technology, 63 pp
24. Novikov D (2013) Theory of control in organizations. Nova Science Publishers, New York, 341 pp

Chapter 4
Analytical Complexity and Errors of Solving Technology Design and Optimization Problems

In this chapter, using the results [1] a uniform search-based estimation procedure for the analytical complexity and errors of solving control problems for organizational and technical systems is presented. It is demonstrated that, first, attempts to reduce the errors cause the rising complexity; second, the complexity goes down as the number of levels in a control hierarchy is increased (under decomposition of control problems); and third, the errors and complexity are natural restrictors for the growth of organizational hierarchies and further application of complex control mechanisms as well as stimulate the choice of typical solutions (patterns).

Organizational and technical systems (OTSs) are characterized, first of all, by a complicated hierarchical structure and a variety of tasks performed at different levels of hierarchy. Therefore, for these systems one often uses the complexes of interconnected control mechanisms, i.e., the mappings of the sets of controlled variables into the sets of control variables (e.g., in incentive mechanisms, these are the relationships between the incentive of an agent and the result of his/her/its activity; in resource allocation mechanisms, the relationships between the amount of an allocated resource and requests for it, etc. [2]). From this viewpoint, as mentioned in the Introduction, a control mechanism is a technology, i.e., a technology of managerial decision-making.

Second, OTSs include agents (individuals, their groups and collectives), which are active and pursue their own goals. Thus, the associated mathematical models proceed from the assumption that the agents seek to maximize their payoff functions, and *hierarchical games* are the basic modeling tool of decision-making and organizational control [3].

As is well-known [4], the maximin principles of optimality, which are typical for hierarchical games, actually suffer from instability. For example, arg $\max_y f(x, y)$ of a continuous function $f(\cdot, \cdot)$ is not continuous, and hence in such problems the first- and second-order oracles [5] cannot be generally applied. (These are standard convex optimization tools employing information on the first- and second-order derivatives of a goal function.) As a rule, numerical optimization in such problems is based on different search methods. The complexity and errors estimates given below are upper

© Springer Nature Switzerland AG 2020
M. V. Belov and D. A. Novikov, *Models of Technologies*, Lecture Notes
in Networks and Systems 86, https://doi.org/10.1007/978-3-030-31084-4_4

bounds: they can be considerably decreased if (in special cases) gradient optimization methods become applicable.

The sets of structurally interconnected optimization problems arising in control problems for OTSs (in particular, in technology design or modernization) should be analyzed in terms of complexity. However, this aspect has received little attention in organizational control [6] so far. The consideration below will begin with general approaches to define and estimate *analytical complexity* (which is comprehended as the worst-case number of oracle launches [5]) and *errors* of solving individual decision problems and control problems for a single agent within the basic model (Sect. 4.1). Then the results will be extended to the case of fan structures, and also the model of growing hierarchies will be studied (Sect. 4.2). Finally, the search for *typical solutions* and *the integration of technologies/mechanisms* in terms of analytical complexity and errors will be discussed (Sects. 4.3 and 4.4, respectively).

4.1   Basic Model

Consider the basic model of an elementary OTS consisting of two active elements—a single control element (*Principal*) and a single controlled element (*agent*). Both elements are active in the sense that each of them pursues individual interests and may have strategic behavior. Let the Principal's goal function $F(x, y, \theta, C)$ be Lipschitzian with a constant L and depend on (a) scalar actions $x \in [0, 1]$ and $y \in [0, 1]$ chosen by the Principal and agent, respectively, (b) the state of the environment $\theta \in [0, 1]$ and (c) a numerical parameter $C \in [0, 1]$. (For the sake definiteness, the l_∞ norm will be used.) In a similar way, let the agent's goal function $f(x, y)$ be Lipschitzian with a constant l and depend on the actions of both elements.

Note that, if a function satisfies the Lipschitz condition, then it is uniformly continuous and hence continuous on the whole definitional domain. The latter property implies that the functions of maximum and minimum are also continuous and Lipschitzian.

Assume the Principal and agent play the *hierarchical game* Γ_2 with side payments. (Recall that the game Γ_2 is a game with a fixed sequence of moves in which the choice of the first player (the Principal who makes the first move) is a function of the action chosen by the second player (agent); see [7] for details.) Then their goal functions take the form $F(x, y, \theta, C) - u(y)$ and $f(x, y) + u(y)$, respectively, where the payment $u(\cdot)$ can be interpreted as an agent's *incentive* given by the Principal.

Denote by $[0, 1]|_h$ the set of points on a uniform grid with a step $h \ll 1$ on the unit segment. Let the function $f(\cdot, \cdot)$ be defined in tabular form using uniform grids with steps H and h in the first and second arguments, respectively (the zero-order oracle in the terminology of [5]). In fact, such a definition corresponds to *uniform search*. In this case, *the analytical complexity W_0 of calculating the agent's best response to an action x chosen by the Principal (the agent's decision model)*

$$BR(x) = \text{Arg} \max_{y \in [0,1]|_h} f(x, y) \tag{4.1}$$

has order $\frac{1}{h}$, i.e., $W_0 \sim \Theta(\frac{1}{h})$. This model yields the maximum value of the agent's goal function with the *error/accuracy* $\Delta_0 \approx \frac{Lh}{2}$ (see the general results in [5]). If the goal functions are incompletely known, the methods described in [4] can be used; or interval optimization methods [8] in the case of the goal functions with inexact coefficients.

Note a couple of important aspects as follows (see the estimates below).

(1) Under sequential optimization (e.g., the summation of maxima/minima), the corresponding complexities and errors have to be added to each other.
(2) Under the iterative calculation of maxima/minima of a certain function, the corresponding complexities have to be multiplied by each other.

As was demonstrated in [4, 6], the Principal's minimum payment

$$u(x, z, y) = \begin{cases} \max\limits_{w \in [0,1]} f(x, w) - f(x, z) + \varepsilon, & y = z, \\ 0, & y \neq z, \end{cases} \tag{4.2}$$

where $\varepsilon > 0$ is an arbitrarily small constant, stimulates the agent to choose the action $y = z$ as the unique maximum point of his/her/its goal function and is also ε-optimal for the Principal. Owing to this fact, the original game can be reduced to the game Γ_1. (Recall that the game Γ_1 is a game with a fixed sequence of moves in which the choice of the first player making the first move does not depend on the actions of the second player [3]). Note that the choice $\varepsilon = lh/2$ in Formula (4.2) compensates the unknown values of the goal function beyond the nodes of this grid; any $\varepsilon > 0$ can be taken if additional information, e.g., on the monotonicity of $f(x, y)$ in y, is available.

Actually, the optimal solution structure (4.2) of the game Γ_2 with side payments [3] eliminates the need for calculating the optimal function $u(\cdot)$; however, this stage is computationally intensive. Moreover, in the case of several agents playing the normal-form game under a given choice of the Principal, Nash equilibrium design generally represents an NP-hard problem [9, 10], and a separate problem here is to find polynomial approximation algorithms for Nash equilibrium (and also to estimate the relationship between the accuracy of such an approximation and the number of agents). For example, see the references in [11].

Assume the function $F(\cdot, \cdot, \cdot, \cdot)$ is defined in tabular form using a uniform grid with steps H, h, p, and q, respectively. In view of (4.2), the calculation procedure for the guaranteed value of the Principal's goal function

$$G(C) = \max_{(x,z) \in [0,1]|_H \times [0,1]|_h} \left[\min_{\theta \in [0,1]|_p} F(x, z, \theta, C) + f(x, z) - \max_{y \in [0,1]|_h} f(x, y) - \varepsilon \right] \tag{4.3}$$

has the analytical complexity $W_1 \sim \Theta((\frac{1}{p} + \frac{1}{h})\frac{1}{h}\frac{1}{H})$; for $\varepsilon = l\,h/2$, it yields the value of (4.3) with the error

$$\Delta_1 \approx \frac{L\max\{h, H, p\} + l(h + 2\max\{h, H\})}{2}.$$

The structure of these expressions suggests to choose the step $p = \max\{h, H\}$, which will be done below.

Under sequential calculations, the analytical complexity can be also used as a rough estimate for the size of required computer memory. For example, with a fixed value C, the calculation of the right-hand side of Formula (4.3) has the complexity W_1, but it is necessary to obtain the relationship $G(C)$ with respect to the parameter C on the uniform grid $[0, 1]|_q$ and the complexity of this operation reaches W_1/q.

Consider the following problem. By an appropriate choice of the steps h and H, minimize the error Δ_1 subject to the constraint T imposed on W_1 in the form

$$(L + 2l)\max\{h, H\} + l\,h \to \min_{h,H}, \tag{4.4}$$

$$\left(\frac{1}{\max\{h, H\}} + \frac{1}{h}\right)\frac{1}{h}\frac{1}{H} \le T. \tag{4.5}$$

This inequality can be interpreted as *the real-time constraint*: given the time τ of one oracle launch, the total time of all calculations constitutes $t = W\tau$. An alternative interpretation is *the limited cognitive capabilities* of a decision-maker.

For the sake of simplicity, the special case $h = H$ will be studied. It follows from the structure of problems (4.4) and (4.5) that the optimal solution makes (4.5) an identity, i.e.,

$$h = H = \sqrt[3]{\frac{2}{T}}. \tag{4.6}$$

For example, for $l = L = 1$ and $T = 10^4$, from (4.6) we have $h = H \sim 0.05$, and also $\Delta_1 \approx 0.1$.

It makes sense to compare the resulting error with the maximum variation of the goal function on its definitional domain. (Recall that this variation is determined by the corresponding Lipschitz constant.) Following this approach, the value $\delta_1 = \Delta_1/L$ can be treated as *the relative error*.

Using a known relationship between the complexity, errors and grid steps, we may pose and solve the error minimization problems (4.4) and (4.5) subject to the complexity constraints and also the complexity minimization problem subject to the error constraints.

Hereinafter, for the basic model let $p = h = H = \sqrt[3]{\frac{2}{T}}$, $W_1 \sim \Theta(\frac{2}{h^3})$ and $\Delta_1 \approx (L + 3l)\,h/2$. In this case, $\delta_1 \sim (1 + 3\,l/L)\,h/2$, i.e., the relative error of Principal's goal function maximization is proportional to the grid step.

The above analysis of the basic model yields a qualitative conclusion as follows. In hierarchical organizational (active) systems, the analytical complexity (as well as the associated computational speed requirements) and also the errors satisfy *the principle of uncertainty*: any attempts to reduce the errors cause the growing complexity; conversely, a smaller complexity leads to higher errors.

4.2   Hierarchical Structures

Consider an OTS of the fan structure. It consists of a single Principal with a goal function $\sum_{i=1}^{n} F_i(x_i, y_i, \theta_i, C_i)$, where the functions $F_i(\cdot)$ are L-Lipschitzian, and also of $n \geq 2$ autonomous agents, which are assigned numbers $i = \overline{1, n}$ (a system with weakly related agents [6]). Because the agents are independent, for each of them the Principal may choose a specific side payment of form (4.2), calculating the corresponding terms $\{G_i(C_i)\}_{i=\overline{1,n}}$ using Formula (4.3).

Under fixed values $\{C_i\}_{i=\overline{1,n}}$, the sequential calculation of $\{G_i(C_i)\}_{i=\overline{1,n}}$ by Formula (4.3) has an analytical complexity of order $\Theta\left(\frac{2n}{h^3}\right)$. Moreover, the relationships between $\{G_i\}_{i=\overline{1,n}}$ and $\{C_i\}_{i=\overline{1,n}}$ have to be memorized, and hence this complexity further raises to $\Theta\left(\frac{2n}{qh^3}\right)$. In *the incentive problem of n independent agents*, the Principal's maximum total payoff

$$G_\Sigma(C_1, \ldots, C_n) = \sum_{i=1}^{n} G_i(C_i) \tag{4.7}$$

is calculated with the error $\Delta_n \approx (L + 3l)nh/2 + \frac{nqL}{2}$ (here the second term corresponds to the solution of problem (4.9) by uniform search). As a result,

$$\delta_n \sim (1 + 3l/L)n\,h/2 + n\,q/2. \tag{4.8}$$

Let us optimize $G_\Sigma(C_1, \ldots, C_n)$, e.g., through an appropriate choice of $\{C_i \geq 0\}$ over the simplex $\sum_{i=1}^{n} C_i = c$, i.e.,

$$g(c) = \max_{\{C_i \geq 0\}:\sum_{i=1}^{n} C_i = c} \sum_{i=1}^{n} G_i(C_i). \tag{4.9}$$

(A possible interpretation of this optimization problem is that *the Principal allocates a resource c (incentives) among the agents*; for the sake of simplicity, let $c = 1$ in (4.9) and (4.13).)

Since the goal function is additive and each term depends on the corresponding variable only, problem (4.9) can be solved using dynamic programming, which has an analytical complexity of order $\Theta\left(\frac{n}{q^2}\right)$. Hence, for the OTSs of the fan structure, the analytical complexity of the entire hierarchy of the optimization problems (4.1, 4.3 and 4.9) has order $\Theta\left(\frac{2n}{qh^3} + \frac{n}{q^2}\right)$. However, as has been mentioned earlier, this chapter considers uniform search as a most universal method that does not utilize the properties of goal functions. For problem (4.9), the complexity of this method can be estimated as $\Theta\left(\frac{1}{q^n}\right)$. We have to calculate maxima (4.3) for each combination of the values $\{C_i\}_{i=\overline{1,n}}$; hence, the analytical complexity of solving the entire hierarchy of the optimization problems (4.1, 4.3 and 4.9) is

$$W_n \sim \Theta\left(\frac{2n}{h^3}\frac{1}{q^n}\right).$$

$$(4.10)$$

Formula (4.8) naturally suggests a series of qualitative conclusions as follows. For the system with a higher number of agents, the error of calculating the optimal value of the Principal's goal function can be retained invariable only by a proportional decrease of the grid step. On the other hand, this causes an exponential growth of the analytical complexity.

Like in many multidimensional optimization problems, the minimization of the error Δ_n (and/or δ_n) is equivalent to the minimization of the grid step h due to their linear relation. However, any attempts to decrease this step lead to a very fast (exponential) growth of the analytical complexity. As an illustrative example, consider the basic model with $l = L = 1$ and the optimal grid steps $h = H \sim 0.05$. For the system with five agents and $q \sim 0.05$, the relative error (4.8) reaches almost 62% while the analytical complexity (4.10) has an order of 2×10^5.

Thus, the errors and complexity are the sources and also natural restrictors of growth in organizational hierarchies; see Sect. 4.3. On the other hand, they stimulate the choice of typical solutions and raise considerable barriers on the way to the "faithful" optimization of the entire organization, top to bottom; see Sects. 4.3 and 4.4.

Up to this point, the Principal's resource allocation problem (4.9) in the fan (two-level) organizational hierarchy has been considered and also estimates (4.8) and (4.10) of the relative errors and complexity, respectively, have been derived. In this context, a natural question is as follows: what will be the errors and complexity if the Principal constructs a three-level hierarchy, partitioning the same set of n noninteracting agents, e.g., into two groups of sizes n_1 and n_2, $n_1 + n_2 = n$, then solves problem (4.9) for each group and seeks for an optimal resource allocation between the two groups?

Denote

$$g_j(c_j) = \max_{\{C_i \geq 0\}: \sum_{i=1}^{n_j} C_i = c_j} \sum_{i=1}^{n_j} G_i(C_i), \quad j = 1, 2.$$

$$(4.11)$$

At the upper hierarchical level, the Principal then solves the optimization problem

$$g_1(c_1) + g_2(c_2) \rightarrow \max_{c_1 \geq 0, c_2 \geq 0, c_1 + c_2 \leq c}. \tag{4.12}$$

By analogy with (4.8) and (4.10), we obtain the following estimates for the errors and complexity:

$$\delta_{n_1 + n_2} = (1 + 3l/L)n_1 h/2 + n_1 q/2 + (1 + 3l/L)n_2 h/2$$
$$+ n_2 q/2 + q/2 = \delta_n + q/2. \tag{4.13}$$

$$W_{n_1 + n_2} \sim \Theta \left(\frac{2n_1}{h^3} \frac{1}{q^{n_1}} + \frac{2n_2}{h^3} \frac{1}{q^{n_2}} \right) \frac{1}{q}. \tag{4.14}$$

In accordance with (4.13), regardless of the agents' partition into groups, a transition from the two-level hierarchy to its three-level counterpart increases the relative error by $q/2$.

Clearly, for $n > 2$ and $n_1 = n_2 = n/2$, it follows that $W_{n_1 + n_2} < W_n$, i.e., by partitioning the agents into two equal-size groups the Principal reduces the complexity.

A more general optimal partition problem can be formulated as follows: under which sizes of two groups (or even for which number of groups) does complexity (4.14) achieve minimum? Denote by m the size of the first group; hence the size of the second group is $n - m$. In a continuous approximation, we have the following binary setup of *the structure design problem*:

$$\left(\frac{m}{q^m} + \frac{n - m}{q^{n-m}} \right) \rightarrow \min_{m \in [0;n]}. \tag{4.15}$$

(Here "binary" means the partition into two groups.)

This setup is elementary: for a given set of agents, generally we may seek for an arbitrary multilevel structure (not necessarily tree) with required properties; see the surveys and results in [6, 12].

The solution of problem (4.15) has the form $m = n/2$. In other words, for reaching the minimum analytical complexity the Principal should partition the set of all agents into two equal-size groups, which reduces the relative complexity by order $q^{n/2}$ in comparison with the fan structure.

Consequently, in the current model with a fixed set of agents, increasing the number of hierarchical levels actually reduces the complexity.

Thus, complexity reduction under an insignificant rise of the errors is a factor stimulating the appearance and further growth of organizational hierarchies. However, this is just one of many other essential factors for a rational choice of organizational structures; see the survey in [12].

Treating the total amount of resource c as a variable, we may pose and solve *resource optimization problems*, e.g., to maximize profits defined as the total payoff minus the total utilization of resource:

$$g(c) - c \to \max_{c \in [0; c\text{max}]} . \tag{4.16}$$

For a given function $g(\cdot)$, problem (4.16) with a grid step h has the analytical complexity $W_c \sim \Theta\left(\frac{c\text{max}}{h}\right)$.

4.3   Typical Solutions

A rational balance between the errors and complexity—the idea described in Sect. 4.2—has been employed there for solving control problems in hierarchical OTSs. However, it can be used in many other fields. Consider two important applications, namely, search for typical solutions (this section) and also the integration of control mechanisms (Sect. 4.4).

Digressing from hierarchical games, let us analyze the agent's decision problem, in which he/she/it chooses an appropriate action in order to maximize the goal function. Recall that the search for the agent's best response (4.1) to the Principal's fixed action $x \in [0, 1]$ has the complexity $W_0 \sim \Theta(\frac{1}{h})$ and error $\Delta_0 \approx \frac{lh}{2}$.

In game theory and decision-making, a widespread concept is the *strategy* of a player or decision-maker (DM). This concept is often defined as a mapping of the set of possible *decision situations* (including the history of a corresponding game, the realized values of uncertain parameters, the strategy profiles of other players, etc.) into the set of his/her/its admissible actions. In the current model, the agent's strategy is the mapping $BR : [0, 1] \to 2^{[0,1]}$ that assigns with the Principal's action (the decision situation for this agent) the set of actions maximizing his/her/its goal function.

For the sake of simplicity, assume the goal functions are such that the agent's best response is unique, i.e., $BR : [0, 1] \to [0, 1]$. If $x \in [0, 1]|_H$, then for each of $1/H$ values of the Principal's action it is necessary to find the agent's action that maximizes his/her/its goal function. The resulting $(1 + [1/H])$-dimensional vector y^*, further called *the complete solution of the agent's decision problem*, is calculated with the analytical complexity $W^* \sim \Theta(\frac{1}{hH})$ and error Δ_0.

Denote by $\{Q_i\}$, $i = \overline{1, k}$, the partition of the unit segment into k connected sets. Let y_i' be the solution of the problem

$$\min_{x \in Q_i|_{H'}} f(x, y) \to \max_{y \in [0;1]|_{h'}} , i = \overline{1, k}. \tag{4.17}$$

The k-dimensional vector y' that is the solution of problem (4.17) will be called *the typical solution*. The idea of using typical solutions (also see the unified solutions of control problems for organizational systems in [6]) can be explained as follows [4]. Instead of the complete set of decision situations (in our case, the unit segment), we take k typical situations in which the agent is suggested to choose corresponding typical solutions (patterns).

For making such typical decisions, the agent has to diagnose the current situation: identify to which of the sets $\{Q_i\}$, $i = \overline{1, k}$, the value x belongs to. (Suppose this problem is solved by the agent without any errors; then the complexity of this procedure—the implementation of a typical solution—can be estimated as $\sim \Theta(k)$). Next, the agent has to choose the corresponding element of the vector y' as his/her/its action. If $k \ll \frac{1}{hH}$, then the complexity of using a typical solution is much lower in comparison with the complete solution of the agent's decision problem.

The error of a typical solution (see Formula 4.17) is $\Delta' \approx \frac{1}{2}\max\{h', H'\}$; for a given partition $\{Q_i\}$, $i = \overline{1, k}$, it has the analytical complexity $W' \sim \Theta(\frac{k}{h'H'})$.

However, a typical solution should be characterized not in terms of the errors but *the price of standardization*, i.e., the goal function losses due to replacing the complete solution with a typical one [1]:

$$\Delta \approx (\{Q_i\}, i = \overline{1, k}) = \max_{i=\overline{1,k}} \max_{x \in Q_i}[\max_{y \in [0;1]} f(x, y) - f(x, y_i')] \approx l \max_{i=\overline{1,k}} \text{diam } Q_i.$$

$$(4.18)$$

Consider the same example with $l = 1$, $h = H = h' = H' \sim 0.05$ ($k = 5$) and the partition into equal-length segments. Then $\Delta_0 \approx 0.025$, $W^* \sim 10^5$, $W' \sim 10^6$ and $\Delta'' \sim 0.2$.

Note that the analysis above has proceeded from the hypothesis that the partition $\{Q_i\}$, $i = \overline{1, k}$, is given. Generally speaking, typical solution design includes two steps as follows: (4.1) find an optimal number k of typical situations (taking into account the DM's cognitive capabilities and other constraints) and (4.2) calculate an optimal partition that minimizes the price of standardization (4.18). The analytical complexity of the latter problem can be very high, especially in the case of multidimensional sets of admissible actions and decision situations.

Thus, typical solutions are justified if they are calculated once but applied many times.

4.4 Integration of Technologies

The book [13] and also the paper [2] considered the integration problem of control mechanisms in OTSs, i.e., the design of *complex mechanisms* that include one or several elementary or other complex mechanisms. The cases of *parallel integration* (a simultaneous and independent application of several mechanisms) and *serial integration* (the output of one mechanism as the input of the other) were described there.

All the control problems studied in Sects. 4.1–4.3 are related with each other (see Table 4.1). Indeed, starting from the agent's decision problem, we have passed to the incentive problem, first for a single agent and then for several independent agents. The established results have allowed us to formulate the resource allocation problem, the structural design problem and the resource optimization problem, one after the

Table 4.1 Relations between technology design/optimization problems

No.	Technology design problem (control mechanism design)	Type of integration [13]	Order of errors	Order of cumulative complexity
1	Agent's decision-making (4.1)	–	$\frac{Lh}{2}$	$\frac{1}{h}$
2	Agent's stimulation (4.2) and (4.3)	Sequential	$(L + 3l)\,h/2$	$\frac{2}{h^3}$
3	Stimulation of n independent agents (4.7)	Parallel	$(L + 3l)\,n\,h/2$	$\frac{2n}{h^3}$
4	Resource allocation (4.9)	Sequential	$(2L + 3l)\,n\,h/2$	$\frac{2n}{h^{n+3}}$
5	Binary structure design (4.12) and (4.15)	Sequential	$((2L + 3l)\,n{+}L)\,h/2$	$\frac{2n}{h^{n/2+3}}$
6	Resource optimization (4.16)	Sequential	$((2L + 3l)\,n + 2L)\,h/2$	$\frac{2n}{h^{n/2+4}}$

other. This sequence of actions is nothing but an integration process of corresponding control mechanisms.

The orders of the analytical complexity and errors for different complex mechanisms are combined in Table 4.1; for the sake of simplicity, all grid steps are set equal to h.

In accordance with this table, both the errors and complexity (especially, the latter) are growing rapidly as we increase the number of mechanisms for integration, in the first place, during transition to optimization in multiagent systems. They exceed all reasonable limits even for a relatively small number of agents. Hence, the errors and complexity are restricting all attempts for a centralized solution of the integration problem of complex control mechanisms. The way out is to use decentralization methods and/or typical solutions and/or heuristic procedures. Of course, each of these approaches reduces the efficiency of control, and the resulting losses should be balanced with possible errors of efficiency optimization under an admissible complexity.

In this chapter, we have suggested an estimation method for the analytical complexity and errors of solving control problems in hierarchical OTSs. The established results can be extended to the following cases:

- the nonuniform grids over convex compact admissible sets;
- the multidimensional actions of system participants;
- the admissible sets differing from the unit segments;
- the agents interacting with each other (based on the decentralization theorems of the agents' game [6]);
- the network structures of OTSs (not trees).

Such extensions seem rather simple and would cause no difficulty of principle.

The conclusion that any attempts to reduce the errors lead to complexity growth (and vice versa, a smaller complexity causes higher errors) is quite natural. At the same time, the decreasing complexity of multilevel hierarchical systems seems to be somewhat unexpected.

The range of problems studied in this chapter belongs to the class C^3 (Control, Computing, Communication); see the surveys in [14, 15]. Really, an explicit analysis of the computational complexity and structure of an OTS, together with the real-time requirements, the cognitive capabilities of an DM as well as other constraints, allows us to compare control mechanisms (in particular, complex ones) in terms of these characteristics and also to perform error optimization subject to complexity constraints.

As has been demonstrated above, the errors and complexity are the natural restrictors of growth in organizational hierarchies and complex control mechanisms. In addition, they stimulate the choice of typical solutions and decentralized approaches (e.g., the ones developed within the framework of multiagent systems and distributed optimization [16–19] as well as algorithmic game theory [10, 20]).

References

1. Novikov D (2018) Analytical complexity and errors of solving control problems for organizational and technical systems. Autom Remote Control 79(5):860–869
2. Novikov D (ed) Mechanism design and management: mathematical methods for smart organizations. Nova Science Publishers, New York, 163 pp
3. Germeier Yu (1976) Non-antagonistic games. D. Reidel Publishing Company, Dordrecht, 331 pp
4. Novikov D (2001) Management of active systems: stability or efficiency. Syst Sci 26(2):85–93
5. Nesterov Yu (2018) Lectures on convex optimization, 2nd edn. Springer, Heidelberg, 589 pp
6. Novikov D (2013) Theory of control in organizations. Nova Science Publishers, New York, 341 pp
7. Ali S, Yue T (2015) U-test: evolving, modelling and testing realistic uncertain behaviours of cyber-physical systems. In: 2015 IEEE 8th International Conference on Software Testing, Verification and Validation (ICST), Graz, pp 1–2
8. Hansen E, Walster G (2004) Global optimization using interval analysis. Marcel Dekker, New York, 728 pp
9. Daskalakis C, Goldberg P, Papadimitriou C (2009) The complexity of computing a nash equilibrium. SIAM J Comp 39(1):195–259 (2009)
10. Mansour Y, Computational game theory. Tel Aviv University, Tel Aviv, 150 pp
11. Czumaj A, Fasoulakis M, Jurdzinski M (2017) Multi-player approximate Nash equilibria. In: Proceedings of the 16th International Conference on Autonomous Agents and Multiagent Systems (AAMAS 2017), San Paolo, May 8–12, pp 1511–1513
12. Novikov D (1999) Mechanisms of functioning of multilevel organizational systems. Control Problems Foundation, Moscow, 150 pp (in Russian)
13. Burkov V, Korgin N, Novikov D (2016) Control mechanisms for organizational-technical systems: problems of integration and decomposition. In: Proceedings of the 1st IFAC Conference on Cyber-Physical & Human-Systems, Florianopolis, Brazil, December 7–9. IFAC-PapersOnline 49(32):1–6

14. Andriesky B, Matveev A, Fradkov A (2010) Control and estimation under information constraints: toward a unified theory of control, computation and communications. Autom Remote Control 71(4):572–633
15. Novikov D (2016) Cybernetics: from past to future. Springer, Heidelberg, 107 pp
16. Boyd S, Parikh N, Chu E et al (2011) Distributed optimization and statistical learning via the alternating direction method of multipliers. Found Trends Mach Learn 3(1):1–122
17. Ren W, Yongcan C (2011) Distributed coordination of multi-agent networks. Springer, London, 310 pp
18. Rzevski G, Skobelev P (2014) Managing complexity. WIT Press, London, 216 pp
19. Shoham Y, Leyton-Brown K (2008) Multiagent systems: algorithmic, game-theoretic, and logical foundations. Cambridge University Press, Cambridge, 504 pp
20. Nisan N, Roughgarden T, Tardos E, Vazirani V (eds) Algorithmic game theory. Cambridge University Press, New York, 778 pp

Conclusions

In this book, an integrated set of models describing the design, adoption and use of complex activity technologies has been presented.

Despite the rigorous considerations at the activity level only (see the Introduction), the results of this book can be efficiently used at other levels of technology modeling. For example, the overwhelming majority of the diffusion-of-innovations models postulate the logistic (S-curve) or bell-shaped dynamics of technology spread, while both curves are a consequence of the corresponding lower-level assumptions; the details have been discussed in Chap. 2. Another example: the results obtained within the framework of technology optimization models (including technological networks) can be used at the subject-matter level for modernizing concrete technologies, etc.

Note that many of the established results allow for a wider range of application, both in terms of their formal models and practical interpretations.

First, historically the learning models have been used for various purposes, not just for technology adoption. A rather simple and general learning model with classical learning curves (exponential, hyperbolic, logistic, and others) as well-interpretable special cases has been successfully constructed in Chap. 2. This fact seems of high epistemological potential for educational, psychological and other studies.

Second, the procedure and results of solving the control problems (see Chap. 3) can be used for other controlled probabilistic processes. The unexpected outcome that the initial uniform probability distribution of system states is minimizing its "asymptotic" entropy requires further comprehension and development.

Third, the results on the analytical complexity and errors of solving some classes of optimization problems (see Chap. 4) are applicable not only to technology design and modernization but also to control design in hierarchical organizational and technical systems. However, the decreasing complexity of multilevel hierarchical systems—an effect that has been discovered in Chap. 4—seems somewhat surprising.

In addition to further theoretical study of the above-mentioned classes of models, a promising line of future investigations is to accumulate typical "technological solutions" with sectoral specifics using the unified general approach suggested in Chap. 2. And possible endeavors to formalize the creative components of technology design are perhaps even more fruitful.

© Springer Nature Switzerland AG 2020
M. V. Belov and D. A. Novikov, *Models of Technologies*, Lecture Notes
in Networks and Systems 86, https://doi.org/10.1007/978-3-030-31084-4

Printed in the United States
By Bookmasters